THE WAR FOR
MUDDY
WATERS

THE WAR FOR MUDDY WATERS

PIRATES, TERRORISTS, TRAFFICKERS, AND MARITIME INSECURITY

JOSHUA TALLIS

Naval Institute Press
Annapolis, Maryland

Naval Institute Press
291 Wood Road
Annapolis, MD 21402

© 2019 by Joshua Tallis
All rights reserved. No part of this book may be reproduced or utilized in any form or by any means, electronic or mechanical, including photocopying and recording, or by any information storage and retrieval system, without permission in writing from the publisher.

Library of Congress Cataloging-in-Publication Data
Names: Tallis, Joshua, date, author.
Title: The war for muddy waters : pirates terrorists traffickers and maritime insecurity / Joshua Tallis.
Description: Annapolis, MD : Naval Institute Press, [2019] | Includes bibliographical references and index.
Identifiers: LCCN 2018052602 | ISBN 9781682474204 (hardback)
Subjects: LCSH: Shipping—Security measures. | National security. | Piracy—Prevention. | Terrorism—Prevention. | Drug traffic—Prevention. | BISAC: POLITICAL SCIENCE / Political Freedom & Security / International Security. | HISTORY / Military / Naval.
Classification: LCC VK203 .T35 2019 | DDC 359/.03—dc23 LC record available at https://lccn.loc.gov/2018052602

♾ Print editions meet the requirements of ANSI/NISO z39.48-1992 (Permanence of Paper).
Printed in the United States of America.

27 26 25 24 23 22 21 20 19   9 8 7 6 5 4 3 2 1
First printing

*To my mom, who always said I would write a book,
and in memory of my father*

Contents

List of Illustrations ix
Acknowledgments xi

Introduction 1

PART ONE Shaping Strategy

CHAPTER ONE *Shifting Tides* 13
   The Threat Forecast 14
   Theorizing Sea Power 21
   Strategy versus Acquisition 29

CHAPTER TWO *Breaking Windows* 37
   The Power of Context 37
   Broken Windows beyond Policing 49
   Room for Debate 52
   Putting It All Together 57
   Approach 59

PART TWO Cocaine and Context in the Caribbean

CHAPTER THREE *The Business of Drugs* 65
   The Region 65
   Narcotics 67
   Counternarcotics 81
   Lessons for Maritime Security 85

CHAPTER FOUR *Trafficking: Guns, People, and Terror* 89
   Guns 89
   People 96
   Terror 109

CHAPTER FIVE *Context and Conclusions*	117
Money Laundering and Misgovernance	118
Lessons from the Coast Guard	126
Lessons Learned	131
Multidimensionality and Context	133

PART THREE Integrating Piracy

CHAPTER SIX *Piracy and Perception in the Gulf of Guinea*	141
The Region	142
Narcotics	144
Life and Crime	150
Piracy	160
Broken Windows	167
CHAPTER SEVEN *Evolving Security Conceptions in the Straits*	175
The Region	175
Piracy	181
Evolution to Broken Windows	192
CHAPTER EIGHT *Charting a Course*	205
Criticism	205
Summation	206
Findings and Consequences	210
Notes	215
Index	253

Illustrations

FIGURE

Figure 1. Contextualizing Maritime Security   5

PHOTOS

USS *Freedom* transits alongside the aircraft carrier
    USS *John C. Stennis*   12
New York Police Department vehicles line Pier 88
    in Manhattan alongside USS *Kearsarge*   36
USS *Zephyr* transits the Caribbean Sea in support
    of Operation Martillo   64
Sailors assigned to Coastal Riverine Group 1 on board
    USNS *Spearhead*   88
Lt. Cdr. Clifford Rutledge with students at Republica
    de Colombia Elementary School   116
Nigerian military forces conduct training on board
    USNS *Spearhead*   140
USS *Coronado* sails with the Royal Malaysian Navy frigate
    KD *Lekiu* and corvette KD *Lekir*   174
USS *Princeton* transits the Singapore Strait   204

Acknowledgments

An incredible number of people devoted time to the fulfillment of this work over many years of effort, which all began, for me, in a small town in Scotland. It was there, at the University of St. Andrews, where I first met with Dr. Peter Lehr, my dissertation supervisor. It was Peter's support—as I navigated developing, researching, and writing the thesis upon which this book is based—that first instilled within me the drive to turn an idea into a PhD and then into a book. It was in that town in Scotland as well where I defended my thesis to my viva examiners (and now friends and mentors), Dr. Timothy Wilson and Dr. Christian Bueger. Their insightful comments, thoughtful recommendations, and continued encouragement were a tremendous boon both for this work and my belief in its future.

In between the start and finish of this project were years of researching, writing, and rewriting, during which I leaned heavily on friends and family. My mother was a constant sounding board, source of support, and partner in celebration. My sisters listened to me talk endlessly about work and lent a helping hand when it was needed. Julian Waller, Jeff Messina, and Katie Winters provided continuous counsel and advice, without which this book would be far worse off, while invaluable proofreading and insight came from many others. I owe a particularly unpayable debt of gratitude and sincerest thanks to my aunt, Mollie Faber, without whose incredible eye for detail, language, and grammar, not to mention enthusiasm and support, this book would have suffered greatly.

Finally, I am grateful to those in my professional world who helped make this project a reality. To Peter Swartz, both for his support in my career and in helping me bring this book to life; to Jim Dolbow at the Naval Institute Press, for seeing its potential and shepherding me through the process; and to a cast of a thousand at NIP working behind the scenes to make this happen, including Rick Russell, Paul Merzlak, Tom Cutler,

Susan Corrado, Emily Bakely, Claire Noble, Robin Noonan, Meagan Szekely, Jacqline Barnes, and freelance copyeditor Patti Bower.

As always, any mistakes, misrepresentations, or just plain bad ideas are my own.

To all who participated in this process with me, please know your time, guidance, and patience were all instrumental in helping turn a dream into an accomplishment.

Introduction

Corrugated tin roofs stretch for miles, dotted and crossed by makeshift markets and roads with no names. Under this urban canopy, countless people subsist on dollars or cents a day. The lucky find work in a sweatshop or factory in the city center, making shirts or toys or shoes for export at pennies an hour. The less fortunate are forced to find work in the teeming illicit black markets found throughout the city's neglected slums. Drug running, human trafficking, weapons smuggling, money laundering, armed robbery, and the myriad petty crimes that accompany these industries become familiar to the world's poorest urban dwellers.

As the sea of fragile metal huts meets the coastline, swarms of creaking wooden and fiberglass dinghies with small outboard motors ferry people, food, and goods on and off shore in a thick stream. There is no discernable municipal presence here, no sign of law enforcement, social services, or sanitation. Life here is dirty and hard, and all too often short. And yet this is life for hundreds of millions of people across the world, from the shantytowns of Mumbai to the favelas of São Paulo. Nearly one-third of the urban population of developing nations, around one billion people, lives in slums like this.[1] As these cities prepare to absorb hundreds of millions more in the coming decades, the world's littorals are fast becoming some of the densest, most dangerous terrain on the planet. How navies meet the threats and crises indigenous to this expanding terrain will in many ways define their relationships with the developing world. Yet the littorals make up a complicated place, a region the U.S. Navy has often preferred to eschew in favor of the familiar security of the open ocean. In this new century, however, it has become obvious that ignoring the perils of the urban coastal poor can be dangerous in global proportions.

❖ ❖ ❖

Late on November 23, 2008, Pakistani terrorists hijacked the Indian fishing trawler *Kuber* while it was out to sea.[2] The captain, Amar Narayan Solanki, had a history of smuggling and likely mistook the terrorists for the usual bands of illicit traffickers who frequent those waters. Solanki may not have resisted his assailants because of his familiarity with coastal smuggling, but his reluctant cooperation would go unrewarded this time. On the evening of November 26, Solanki was bound and killed as the terrorists disembarked to land near one of the many slums on the outskirts of Mumbai. Their presence did not go unnoticed; locals spotted both teams as soon as they grounded their inflatable boats near Badhwar Park and Machhimar Nagar. While one raiding party passed itself off as a group of students, the other was able to successfully intimidate the residents into remaining quiet, likely because the residents (as had Solanki) assumed them to be part of the fabric of political violence, including organized crime and smuggling, that defined their lives. The terrorists who had slipped out of Pakistan amid the traffic of coastal slums had infiltrated India in much the same way, easily "inserting themselves into a coastal fishing fleet to cover their approach."[3]

Yet the havoc they caused was exponentially greater than that of Mumbai's typical smuggler. One of the world's largest cities was brought to a paralyzed halt as more than one hundred people were killed and hundreds more wounded in one of the most shocking and successful terrorist attacks in modern history. The terrorists, all but one of whom died in the attack, were at home in the urban camouflage of Mumbai's slums. And while the result was unmistakably an act of terrorism, the method of its execution "blurred the distinction between crime and war: both the Indian ship captain and the local inhabitants initially mistook [the terrorists] for smugglers, and their opponents for much of the raid were police, not soldiers."[4] The landing sites these terrorists chose are illustrative of the littoral terrain that navies so frequently hope to avoid, littered with "dense, complex informal settlements—coastal slums made up of thousands of tiny shacks, fishing huts, and moored boats."[5] To Solanki and the residents of these slums, there was nothing out of the ordinary about those strange men landing on their shores; it was part of the ebb and flow of life on the margins. Yet the littorals are bringing the margins to the foreground, and understanding this terrain is central to understanding its impact on the broader global system.

❖ ❖ ❖

The famed British naval strategist Julian Corbett notes simply, "Men live upon the land and not upon the sea," and consequently a navy's real task is to influence events ashore.[6] Naval historian Capt. Dudley Knox, a contemporary of Corbett, echoes the same principle: "The supreme test of the naval strategist is the depth of his comprehension of the intimate relationship between sea power and land power, and of the truth that basically all effort afloat should be directed at an effect ashore."[7] There is no easy definition for "littoral," but the spirit of these quotes reveals the principle of the word, the region where sea power and land power overlap.

Broadly speaking, the littoral region is the "area of land susceptible to military influence from the sea, and the sea area susceptible to influence from the land."[8] In military terms, "a littoral zone is the portion of land space that can be engaged using sea-based weapon systems, plus the adjacent sea space (surface and subsurface) that can be engaged using land-based weapon systems, and the surrounding airspace and cyberspace."[9] The littoral is therefore defined by the technological capabilities of a military, and as a result the littoral is not like other geographic terms. A mountain remains a mountain no matter the century; a plateau and a river and a steppe do not require points of technological reference. Yet time has changed what we mean by littoral.

In an era of cannons, large, shore-based batteries determined the distance of the littoral to be as far as a favorable wind could carry a projectile (about three miles, the original extent of the territorial sea).[10] In the U.S. war in Afghanistan, by contrast, the U.S. Navy demonstrated that it could contribute to operations hundreds of miles inland, radically changing what we mean by littoral. If we conservatively limit the littoral to just two hundred miles inland, that would encompass 75 percent of the world's population, 80 percent of capital cities, and practically "all major centres of international trade and military power."[11]

The littorals are a congested environment, crowded by freight in transshipment, oil and gas platforms, navigational markers, and geographic features like islands and reefs. The hive of human activity in the region only serves to constrict movement even further, with littoral zones cluttered by cargo ships, aircraft, and leisure and small commercial vessels. Furthermore, time has not only expanded the littorals horizontally. Today navies must consider up to nine operational domains:

> The seabed, the submarine environment, the sea surface, and naval airspace (airspace over the sea), which together make up the maritime domain; the land surface, subterranean space, and supersurface space (to include tunnel systems, canals, sewers, basements, exterior street-level surfaces, building interiors, high-rise structures, and rooftops), making up the land domain; the airspace domain; and the domain of cyberspace.[12]

In other words, the littorals are where the threats are, where responses must be rapid, and where nonstate and nontraditional actors can have an outsized impact on the global stage. How navies will recalibrate to meet the growing threats of sprawling littorals will be central to defining their impact on the global system in the coming half century.

In this book, I won't get overly technical about defining the littorals. For reference, we can simply use a state's exclusive economic zone (measured as two hundred nautical miles perpendicular from the coast as set forth in the UN Convention on the Law of the Sea) and the adjacent land as baselines. The goal is to spark a conversation on the distinct asymmetric actors and insecurities plaguing this ambiguous domain—the very actors and actions that straddle the lines of convention. And ultimately, discussing the littoral zone requires a comfort with ambiguity. Time and tide change the exact definition of the terrain from day to day, yet its fundamental role in human security and society remains constant.

❖ ❖ ❖

It is also incumbent upon us to narrow our definition of the term "maritime security." The emerging field of maritime security, even more so than similar disciplines like terrorism studies, has yet to settle on a universal definition of the term. This is so in large part because, as Christian Bueger of Cardiff University illustrates, maritime security in its modern usage can be portrayed at the nexus of a range of overlapping spheres of security (see figure 1).

This approach to the definitional issue is semiotic, an attempt to understand a concept in relation to known others. Yet one of the greatest challenges with such an approach rests in achieving a narrow enough definition to ensure fruitful dialogue. Until we can accomplish this much, as both Bueger and Basil Germond (of Lancaster University) note, the concept of maritime security will remain more of a "buzzword" than a useful tool.[13]

King's College London's Geoffrey Till regards maritime security's buzzword quality and ambiguity as particularly problematic, regarding the concept as an overcomplicated rebranding of the more conventional idea of "good order at sea" (itself not a particularly clear phrase, in my opinion).[14] Indeed, as Till notes, the word "maritime" alone connotes a range of meanings, from entirely military to purely commercial.[15] Still, while we cannot yet point to a consensus definition of maritime security (and, truly, in few security fields does such consensus exist—take terrorism studies), we can nevertheless embark on a worthwhile conversation on this issue so long as we employ a consistent and common usage within our own bodies of work.

In that spirit, here maritime security is the state response to threats precipitated by nonstate, frequently transnational actors who, while operating, break national and international law and maintain the capacity to use violence in the furtherance of such actions. In other words, "maritime security has to do with (illegal and disruptive) human activities in the maritime" domain.[16] This trims down some of the subcategories you might find on a matrix of "maritime issues" (see Figure 1), like excluding marine safety issues (under the "Marine Environment" tab) or the interstate conflict element of the national security category.[17] We are still left with a term at the nexus of multiple conceptions of security but one aimed at a single framing of the nature of the threat (asymmetric, illegitimate) in much the way traditional naval strategy is often conceived in terms of one major framework (symmetric, legitimate, interstate threats). This is consistent with Germond's definition of maritime security, which emphasizes the restoration or maintenance of the legitimate use of force by state actors.

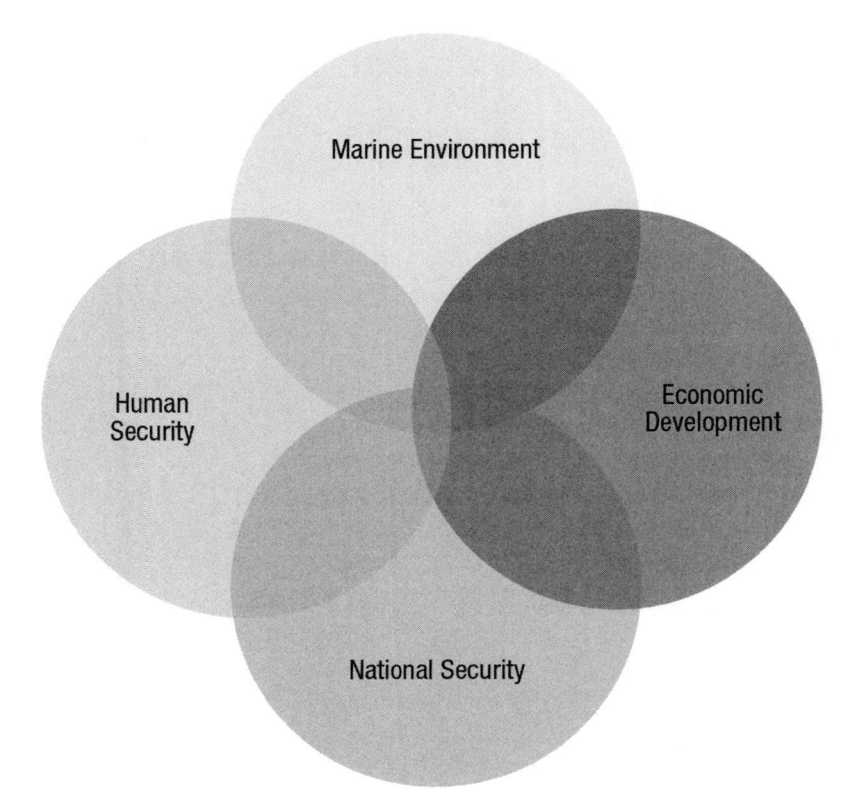

Figure 1. Contextualizing Maritime Security. *Derived from Christian Bueger, "What Is Maritime Security?,"* Marine Policy *53 (2015): 161.*

Maritime security, when defined like this, has both geographic and political dimensions. If we deconstruct the phrase, *maritime* speaks to the geography of the domain (the littorals) while *security* speaks to the nature of the threat (illegitimate nonstate actors). Maritime is also significant as it stands apart from more conventional conversations on "naval" strategy. The word "maritime" is evocative of the broader sea-based international system, the role oceans and rivers play in the geopolitical as well as geoeconomic interactions of countries across the globe. "Naval" speaks principally to the actions of navies and is laden with centuries of historical connotations. In this book, when I use the word "naval" (or "naval strategy"), it is in reference to this specific heritage of conventional naval theorizing, one historically focused on interstate conflict.

❖ ❖ ❖

U.S. interests in the stability of coastal zones worldwide face a diversity of threats, many of which arise from nonstate and transnational actors. These regions are hotspots for the trafficking of people, narcotics, and arms, operations that destabilize friendly governments and support terrorist activities. These transnational organizations and global movements of people and goods over the coming decades could pose some of the most prolonged (even if not existential) threats to U.S. interests and global stability. While this in no way precludes the rise or significance of other traditional threats, the unconventional nature of maritime insecurity poses an unusually fundamental identity crisis for where (and how) the U.S. Navy operates.

While maritime insecurity takes place on the water, the literature explored throughout this book reflects threats that often seem to have more in common with crime than conventional naval warfare. The exposition above reads nothing like the imaginings of Alfred Thayer Mahan or Corbett, the canonical fathers of naval strategy (more on them in chapter 1). And so the question arises: can any of the research done on crime help us understand maritime issues in a new way? How about a specific theory of crime—could that help us build a theory of maritime security? To be more precise, can the broken windows theory, an originator of community policing theories, help us construct a strategy for maritime security?

But how can we take an idea from community policing and make it relevant for maritime security? We can start by finding the right lessons to export. These lessons will be derived from two areas of exploration, one theoretical, one real world. In other words, we can mine both what the academics tell us about community policing and how the theory works in practice to find the best building blocks for a maritime security strategy.

Notably, I am not among those criminology academics—my expertise is in maritime security and political science. Those with mastery of that field will likely find wide gaps in my knowledge. I ask forgiveness for straying across party lines and remain optimistic that the larger intellectual effort represented by this book stands. That is, that maritime security has a lot to learn from the world of crime.

As a theory, broken windows holds that our actions are subconsciously influenced by our environment, both the physical space we inhabit and how we feel about that space. In other words, what economists might call the rational actor model—a world in which we make the best possible decision based on as much information as we can get—cannot fully explain human behavior (including crime). We are all constantly responding to subtle stimuli from our physical and behavioral context. Broken windows, as a theory, suggests that we think about how police (or, in our case, maritime forces) might use that knowledge to help build security.

In practice, when police tried to focus on the environmental signals that influence behavior, they found that reality was more complicated. As it turns out, low-level criminality is highly multidimensional; in other words, policing small-scale disorder (like graffiti or turnstile hopping at the subway) has an outsized impact on the rates of more violent crimes. This can, of course, drive police to unhelpful extremes, criminalizing acts of petty crime beyond sensibility (we are wrestling, as a nation, with that fallout today). But in measured implementation, police have found that when they try to alleviate crime by tackling the signals of disorder in our environment (like real broken windows), doing so indirectly tackles crime along a wider continuum.

Okay, but how do you connect this conversation on community policing and the broken windows theory to maritime security? To do that, I advance a strategy-building argument in three sections—Shaping Strategy, Cocaine and Context in the Caribbean, and Integrating Piracy. Part 1, Shaping Strategy, speaks to relevant naval and criminological theories upon which this book rests. Chapter 1 lays the groundwork for understanding the naval context. The chapter outlines the global trends that make maritime security a pressing issue for the U.S. Navy and how these trends are, and are not, addressed in some of the existing strategic literature. Chapter 1 concludes by bringing this conceptual debate into the real world, exploring persistent challenges with which maritime security strategists have to contend.

Chapter 2 turns to policing, explaining the broken windows theory and identifying key themes to which we will return throughout the book.

In particular, I explore how I got to the two ideas central to broken windows policing referenced above—that crime is informed by context and that such activity is multidimensional. Context, signals from our environment, influences behavior in subtle but important ways. Multidimensionality, meanwhile, suggests that individual acts of disorder are part of a wider network of criminality, often in unexpected ways. This chapter also explores in what ways the broken windows theory has been employed outside of criminology, which is important in demonstrating that the theory can be valuably exported to new contexts. The final section of this chapter widens the aperture one last time. It details how the broken windows theory relates to a wider set of security and maritime security ideas.

Part 2, Cocaine and Context in the Caribbean, provides a proof of concept for how the ideas drawn out in the first section connect to the maritime space in practice. This section relies heavily on operations and documents from the U.S. Coast Guard. With its constabulary responsibilities, the Coast Guard provides an ideal bridge for exploring how policing techniques might be brought to bear in a naval context. Thus, while the purpose of this book remains to better inform wider U.S. naval strategy, there is an undeniable and integral relationship with the Coast Guard's experience in the Western Hemisphere. Across three chapters, I employ a detailed portrait of maritime crime in the Caribbean in order to demonstrate how a broken windows approach to thinking about maritime security proves relevant and instructive. Beginning with narcotics trafficking in chapter 3, I explore how community policing helps reframe an enduring issue in a new way. Chapter 4 expands that framing to assess a broader set of threats, like human trafficking and terrorism, using a unified strategic lens. Finally, in chapter 5, I incorporate less visible issues such as corruption and money laundering into this strategic framework. By the end of this section, you'll see how looking at the Caribbean through the perspective of policing provides the foundation for a new theory of maritime security.

Leveraging the proof of concept from the second section, we can put this new theory to the test in part 3, Integrating Piracy. I explore this theory testing in two chapters, one on the Gulf of Guinea and one on the Straits of Malacca and Singapore. Chapter 6 begins by showing that some important conditions sketched in the Caribbean—that crime is context-dependent and multidimensional—are reflected in West Africa. As a consequence, the theory employed in the Caribbean should also be valid in the Gulf of Guinea. If that is in fact the case—if the theory is generalizable—it should be able to incorporate new information (or new threats) and retain its usefulness. So I put that to the test, with piracy. Successfully introducing a new issue in a new region (piracy in West Africa) would show that the connections between policing and maritime security are not unique to the Caribbean or its challenges. Chapter 7 takes this conversation on piracy one step further, exploring it in a new context (Southeast Asia), one that has wrestled with pirates for decades. In this final case, I demonstrate why a new theory on maritime security is useful even in a region that maritime security scholars have been looking at for a long time. If here, in this context, the theory has something useful to say, that goes a long way to making a meaningful contribution to a strategy of maritime security.

Finally, the last chapter translates this strategic and theoretical story into actionable recommendations for strategists and researchers. This section articulates specifically how the U.S. Navy—and, indeed, any maritime force—might employ a broken windows theory of maritime security to better improve decision making and prepare for a world in which maritime insecurity is playing an increasingly prominent role.

PART ONE
Shaping Strategy

The littoral combat ship USS *Freedom* (LCS 1) transits alongside the aircraft carrier USS *John C. Stennis* (CVN 74) during a group sail exercise. The U.S. Navy ships are under way conducting an independent deployer certification exercise off the coast of Southern California. *U.S. Navy photo by Mass Communication Specialist Seaman Kenneth Rodriguez Santiago/Released*

CHAPTER ONE
Shifting Tides

"**M**aritime security" is an increasingly common subheading in American strategic documents. Yet, while the use of the phrase has expanded, there remains heated debate about how maritime security operations (principally in the littorals) fit into the wider realm of U.S. naval strategy. It is no stretch to say that the U.S. Navy, as an institution, is more interested in and concerned about blue water threats than those emerging from nonstate actors in murkier waters. And yet, as I explore in this chapter, historical trends, emerging patterns, and a burgeoning strategic literature conspire to bring the littorals to the forefront. These forces suggest that maritime security—naval operations in green and brown waters directed against nonstate or unconventional actors—will play an increasingly important role in U.S. foreign policy.

In this chapter, I endeavor to explain why maritime security, as a still refining field of study, warrants increased theoretical attention and investment from naval strategists. This argument is addressed in two parts. The first, The Threat Forecast, investigates the emerging predominance of the littorals, providing an understanding of the threats the Navy may have to weather over the coming decades. In the second section, Theorizing Sea Power, I explore the timeline and content of select post–Cold War, largely post-9/11 naval literature to help place into context the evolution and growth of maritime security strategy. I conclude that such trends, both real world and theoretical, will continue to draw the Navy toward low-intensity littoral operations in the decades to come—whether it wants to or not. Notwithstanding that draw, in a final third section, Strategy versus Acquisition, I highlight a persistent gulf between the recognition of the strategic challenges the littorals pose and the resourcing necessary to combat them.

The Threat Forecast

Why should we care about maritime security at all? "The United States faces a rapidly changing security environment," answered the 2014 *Quadrennial Defense Review* (*QDR*) in its opening lines. From "new centers of power" to "a world that is growing more volatile" to "unprecedented levels of global connectedness," the nature of military strategy is being challenged at numerous points.[1] The 2014 *QDR*'s predecessor approach was, as the report notes, "fundamentally a wartime strategy." Yet the strategic white papers emanating out of the United States in the mid-2010s stressed that as the United States and its allies attempt to move past these wars, what comes next will be no less ambiguous. Conflicts of the future may be characterized by hybrid threats from nonstate, asymmetric actors to conventional, symmetrical adversaries armed with nuclear weapons. The dynamic nature of littorals and the increasing interconnectedness of global commerce and information flows demand that theorists revisit what we mean by maritime security.

There are a growing number of unconventional factors that are poised to require greater consideration from navies in the coming decades. In his book *Out of the Mountains*, counterinsurgency strategist David Kilcullen sets out a compelling argument for taking the littorals seriously. While Kilcullen's argument is focused on events ashore, his articulation of the global drivers shaping the littorals is equally valuable for the seaward end of the domain. The premise underlying these drivers is based on one simple principle: conflict happens where people are.[2] So, where are the people?

To answer that simple question, Kilcullen identifies four "megatrends" of demography and economic geography that suggest where we will find most of the world's population in the coming decades. "Rapid population growth, accelerating urbanization, littoralization (the tendency for things to cluster on the coastlines), and increasing connectedness" all suggest that populations are concentrating in networked, urban, dense, littoral communities.[3]

UN estimates project that by 2050 at least two-thirds of the global population will likely live in cities, a substantial portion of the anticipated 9.8 billion global inhabitants by that date. A massive proportion of urban dwellers already live in slums (about a billion people, as noted in the introduction)—a figure that is only likely to rise in real terms and maybe even as a percentage of the urban population. Already nearly 1.5 million people migrate to a city *every week*, and cities are soon expected to absorb almost all new population growth.[4] Moreover, most of this urban coastal expansion will take place across the Global South, often where governments are least prepared. The populations of the Caribbean and Latin America could grow by more than 130 million people; Africa may see its population expand by more than a billion; and Asia will see a growth of at least 800 million (all mean estimates from the United Nations).

In the 1990s the World Bank was already anticipating that slums would become the "most significant, and politically explosive, problem of the next century."[5] To give an indication of scale, consider the following. In one generation, the world's poorest cities will absorb most of a population increase equal to all the population growth ever recorded in human history through the year 1960.[6] That is sure to produce astounding institutional strains, ones that will inevitably stress already fragile municipal governments to a breaking point. And yet, only since around 2006 and the "Trends and Shocks" project at the Department of Defense does it appear that demographic projections have played a central role in the strategic planning process.[7] And we need not wait for 2050 to see these statistics in action. Already half of the world's population lives within about thirty miles of a coast, and three-quarters of large cities are on the water.

But how is demography dangerous? The answer comes from how communities respond to the pressures of these trends. Research suggests that when a city doubles in size, it produces, on average, "fifteen percent higher wages, fifteen percent more fancy restaurants, but also fifteen percent more [HIV/AIDS] cases, and fifteen percent more violent crime. *Everything* scales up by fifteen percent when you double the size."[8] In cities already bursting at the seams, cities like Lagos or Mumbai or San Pedro Sula, the consequences of a 15 percent spike in crime or HIV/AIDS rates act on preexisting stressors like poverty, climate change, and political violence, which can precipitate still more disorder. In the most extreme, the impact of these magnified stressors might even cause a metropolis to turn "feral." In the *Naval War College Review*, Richard Norton defines a feral city as "a metropolis with a population of more than a million people in a state the government of which has lost the ability to maintain the rule of law within the city's boundaries yet remains a functioning actor in the greater international system."[9] John Sullivan and Adam Elkus describe the feral city as a place without any meaningful presence of legitimate authority, where "the architecture consists entirely of slums, and power is a complex process negotiated through violence by differing factions."[10] Norton noted at the time that only Mogadishu truly exemplified his criteria of a feral city. Yet, several cities then contained feral pockets characteristic of Norton's theory. In such pockets, segments of an urban or peri-urban sprawl (a region at the intersection of urban and rural spaces) exist outside the reach of the government. A *Proceedings* article by Matthew Frick as well as a *New York Times* piece both cite São Paulo, Mexico City, and Johannesburg (the *Times* also adds Karachi) as examples of modern cities with feral components.[11] Frick explicitly links these feral cities to maritime security by arguing that they are modern echoes of the pirate havens of lore, inheritors of the piratical legacy of leveraging ungoverned spaces as bases of operation. In feral cities, the legitimate governing fabric of the urban area erodes as the stress of an oversized population pushes it past the ability to cope with the pressures of crime, poverty, and health. This breakdown perpetuates the cycle of urban exclusion, precipitating yet more physical, political, social, economic, and infrastructure neglect.[12]

The collapse or erosion of local governance therefore creates a power and services vacuum. That power vacuum creates disorder from which people crave a reprieve. In turn, the local actor that can produce a sense of order and stability is frequently adopted as a surrogate government. Inevitably, these surrogates are not peaceful neighborhood watch groups. Kilcullen's concept of "competitive control" posits that the surrogates most likely to survive in these environments are those that can act across a spectrum of power ranging from soft to coercive.[13] Without soft power, an institutionalized set of normative community rules, the population is denied the sense of order it demands. Under such a rules-based system, a neighborhood is likely to tolerate a measure of violence because it can be predicted and avoided. Without violence, a competitor will find it easy to supplant the surrogate. What we are left with is a void filled by organized gangs, terrorists, militias, or criminal networks that manipulate a feral city's disorder to establish fortifications within neighborhoods that then often come to depend on them. Kingston, Jamaica's garrison neighborhoods offer one such example, districts made loyal to gang leaders because of their dominant role in the local informal justice system and economy.[14] When such groups offer locals enough of a sense of fair and predictable order, many are willing to tolerate their presence.

One of the most sophisticated and successful examples of the rise of these surrogate groups comes from Hezbollah in Lebanon. The group is capable of acting across the range of Kilcullen's spectrum, from responsive normative governance structures to armed coercive capacity.[15] Hezbollah has opened hospitals and schools and has provided job training and security to Lebanon's impoverished Shia community. Filling an expansive power vacuum, the organization's ability to operate everything from daycares to militia units, its subsequent subnational hold on southern Lebanon, and the

2006 war it precipitated with Israel are all proof of how entrenched and dangerous nonstate actors can be. At its root, we are talking about state capacity (or lack thereof) to monopolize the use of legitimate violence, the essence of sovereignty as described by Max Weber.

So far, however, surrogate groups and feral cities would appear a danger largely within their own neighborhoods, far from concerning the world's largest navy. What brings these groups and populations into the international maritime orbit is Kilcullen's final megatrend—connectedness. Crowded, poor coastal zones may turn feral, but feral regions do not implode. Instead, they remain connected with the world around them through the Internet, ports, airports, and diaspora communities. Thus, with all the implications of such connectedness, even a feral region remains a dynamic, strategic, and significant actor in the international arena. This connectedness allows the feral elements of a community to interact with licit and illicit trade and information flows, offering local actors the capacity to directly impact events across the country and world.[16] The same pathways that facilitate normal trade and migration likewise enable illicit transfers such as the smuggling of narcotics, people, and weapons; human trafficking; terrorism; and piracy. This connectedness is what gives rise to the potential for actions a world away, events like protests or acts of political violence, to have a rapid and meaningful consequence for domestic considerations elsewhere (something the 2014 *QDR* recognizes, by the way). Because of the connection between local nonstate groups and international security, "the distinction between war and crime, between domestic and international affairs" almost disappears.[17] We are left facing an amorphous hybrid threat, what John Sullivan and Adam Elkus call "criminal insurgents," or the melding of criminal syndicates with the direct political control of territory.[18]

Military theorists have often resisted addressing hybrid threats, falling victim to a bureaucratized approach to security that categorizes (perhaps even "curtly dismisses") it as an issue of law enforcement.[19] That seems to have broken loose somewhat over the past several years, driven in part by the actions of Russia in Crimea. We can see why from retired lieutenant colonel Frank Hoffman's definition of a hybrid threat, which he describes as "any adversary that simultaneously and adaptively employs a fused mix of conventional weapons, irregular tactics, terrorism and criminal behavior in the battle space to obtain their political objectives."[20] While Hoffman's definition does not specify the nature of an adversary, over the years since Russia's annexation of Crimea the discussion of hybrid warfare has gained steam with an eye toward how national actors leverage unconventional tactics to achieve classical objectives (thereby getting muddled with the Gray Zone debate as well).

Here I'm using the same tactical definition Hoffman employs but to explore how nonstate groups move between threat categories, mixing terrorism, crime, insurgencies, and the like. This nonstate focus is represented in much of the strategic literature from the post-Iraq transition era. The 2010 *National Security Strategy* includes a section titled "Transnational Criminal Threats and Threats to Governance," for example. In it the Barack Obama administration notes that illicit trades and the transnational actors who facilitate them continue to "expand dramatically in size, scope, and influence," and that such groups have "accumulated unprecedented wealth and power."[21] Thus, while the concept of the interchange between local and transnational issues, between issues of criminality and issues of larger security consequence, has struggled to come into a clear and distinct phraseology, the growing influence of such threats already commands attention. Demographic trends further portend that hybridized threats may soon impact hundreds of millions more people in the coming decades. The surrogate groups likely to produce these threats are diverse, including gangs, insurgents, terrorists, vigilante groups, and criminals of a wide variety (including pirates and traffickers).[22] All remain connected to the wider world through the littorals.

A hallmark of hybrid threats is that there is a great deal of overlap and confusion between crime and war, and therefore a greater blurring of the lines between what police, constabularies, and militaries do.[23] Consider the constabularization of some military or expeditionary units to meet these complex demands. Australia, for example, fields an internationally deployable federal police unit, and the Italian Carabinieri host the North Atlantic Treaty Organization (NATO) Center of Excellence for Stability Policing Units, which trains students in community policing methods.[24] Israeli police form joint operations forces with the military to help close the gap between operational requirements.[25] The U.S. military studies policing methods and has included police trainers on deployments abroad.[26] These forces do so because hybrid (frequently littoral) threats are distinct. And they are not new. On land there exists a "long-standing historical pattern in which the United States conducts a large-scale or long-duration counterinsurgency or stabilization operation about once a generation and a small or short-term mission about once every five to ten years."[27] In truth, the American military faces nontraditional challenges more frequently than it does interstate conflicts. As the line blurs between war and crime, it may only become more attractive for strategists to circumvent planning for these opaque, nontraditional challenges in favor of conventional considerations. History, however, suggests that would be a mistake.

What is remarkable about the pattern of unconventional conflict, as Kilcullen outlines well, is its irreverence for the preferences of policymakers.[28] Lyndon Johnson, who placed a premium on domestic reform, was eventually subsumed by a troop escalation in Vietnam. Bill Clinton delayed sending troops to the Balkans and resisted doing so in Rwanda yet ultimately sent troops to Bosnia, Kosovo, Macedonia, Haiti, and Liberia over the course of his presidency. While candidate George W. Bush dismissed any interest in stability operations and nation building, President Bush launched two simultaneous large-scale counterinsurgency campaigns and led the United States into the largest campaign of nation building since World War II and NATO into its largest stabilization mission in history.[29] Even as President Obama initiated his pivot to Asia, he was still negotiating the drawdown of two land wars while conflicts simmered (and flared), inter alia, in Syria, Yemen, Somalia, Iraq, Mali, Nigeria, and the Congo. This is not strictly a contemporary phenomenon either, the meeting of large militaries with small conflicts. The British Army, for example, engaged in myriad low-intensity brushfire wars across its empire. The American military "has been drawn into literally dozens of small wars and irregular operations" since its inception, with its formative years defined by countless skirmishes with both the British and Native Americans.[30] Even large conventional wars have included substantial nontraditional components and guerilla tactics (consider the partisans in World War II).[31] "American policy makers clearly don't like irregular operations, and the U.S. military isn't much interested in them, either," in Kilcullen's estimation, but they will nevertheless continue to happen, and perhaps at increasing frequency according to demographic trends.[32]

The frequency of these operations drives Kilcullen to reevaluate some of the very counterinsurgency theories he helped popularize, including the fundamental principle of "winning hearts and minds." The idea that communities can be won over by appealing exclusively to their rational interests ignores the fact, as the broken windows theory articulates (see the following chapter for more), that behavior and environment dynamically shape one another. The primary project is not defeating insurgents or even winning hearts and minds. "The project is the community"; the project is context and environment.[33]

Focusing on context, however, is not so clear-cut. As broken windows practitioners came to understand, shaping context requires distinguishing between operational and strategic success. Brazil's military policing unit (BOPE) is singularly effective at combating cartels and gangs on a tactical level, for example. Yet the methods they have employed may alienate the poor or lock favelas into perpetual perceptions of conflict.[34] The Israel Defense Force (IDF) has perfected a method of using the city as a medium for conflict, moving through walls (blasting holes between buildings) in an effort to deny combatants a conventional theory of maneuver.[35] And yet such tactics terrify noncombatants and leave the urban space in tatters. While the BOPE and IDF are operationally adept, neither Latin American organized crime nor Hamas terrorists have been dislodged from their strongholds. This is, in some very small part, because both the BOPE and IDF neglect the power of context. Thus, as Sullivan and Elkus write, while these forces "may be masters at manipulating the city as fluid operational space, they have an overly materialistic conception of the city," ignoring the people in deference to surmounting tactical maneuver challenges. In other words, these forces "ignore the *social spaces* in their strategies."[36] In the social arena, militaries have a great deal to learn from police.

The connections between hybrid threats, police, and littoral security are being increasingly recognized. Kilcullen, for one, asks, "What if we could combine what I learned in Baghdad about protecting urban populations from extreme violence with what law enforcement agencies know about community-based policing?"[37] While he was focused on the land, the same question applies to both halves of the littorals. And the answer, as we see throughout this book, is that effectively combating maritime threats means meeting the needs of the communities from which they spring—and on which they prey. Operating in littoral environments with a sole focus on the threat, and without regard for the wider community, can produce overly restrictive approaches that choke the life out of places, a phenomenon sometimes called "urbicide" with respect to cities. Such a community may be pacified, but the region neither serves nor enriches its inhabitants. Blunt force tactics produce "military urbanism," what results when militarized enforcement "turns cities into fortresses and populations into denizens of occupied territory."[38] To achieve durable stability in the littorals, we need a strategy that can blend the techniques and knowledge of police with the matériel, scale, and operational sophistication of maritime forces. In other words, what is needed is a strategy of maritime security.

It is logical to conclude that hybrid threats are best met with a hybrid strategy, one that equally blends the capabilities of maritime forces with the theories of policing. Such a strategy, anchored in both naval and criminological roots, would be more effective against hybridized threats, on the one hand mobilizing greater capabilities than police possess while on the other wielding that force through less destructive tactical approaches than those sometimes employed by militaries. How to find that blend is a major theme of the chapters to come.

At the start of this chapter I noted that this first half would explore the maritime threat forecast, pulling as we did on Kilcullen's megatrends, demographic factors, and an honest accounting of the predominance of unconventional conflict in recent history (whether sought out by policymakers or not). It should be noted, however, that any attempt to predict the exact nature of a future threat or conflict is a fool's errand. As a consequence, it is important to understand how to think about the factors introduced in brief above as well as what a threat forecast is supposed to tell us. To borrow a metaphor from Kilcullen, much like how climate modeling cannot say whether it will rain or snow next week, the demographic projections above say little about the near-term conflict forecast. Yet, just as with climate models, such forecasts "do suggest a range of conditions—a set of system parameters, or a 'conflict climate'—within which [future] wars will arise."[39] These projections illustrate trends. And while those trends do not say much about what happens tomorrow or the day after, they speak volumes about the forces steadily reshaping our world. These forecasts are unequivocally telling us that dense, networked, and littoral communities are an emerging global force. How maritime forces meet (or fail to meet) the challenges associated with that rise will be dictated by their attentiveness to the unique constraints of securing muddy waters.

THEORIZING SEA POWER

If, as I argue, much of the future lies in securing the littorals against unconventional threats, why does there appear to have been so little by way of conceptual (theoretical and strategic) interest in them by the U.S. Navy? The answer may lie, in part, in bureaucracy: both in how organizations competing for resources fall back on trusted ideas and in how academia may feed a bureaucratic tendency to silo problem solving.

First, at the broader historical level, one interesting answer may rest in the theory that strategy serves an important function as a cudgel of bureaucratic self-defense for the services. To demonstrate this point, University of Michigan professor John Shy identifies three historically significant American military strategies: sea power, strategic air attack, and limited war.[40] In each case, Shy argues, these strategies were just as much products of the military that employed them as they were responses to the needs of their time. These strategies are legacies of an impulse toward self-justification, according to this lens, each a response to a branch's crisis of legitimacy: in the Navy in the 1880s, in the Air Corps in the 1920s, and in the Army directly following the close of World War II. Shy suggests that each of these strategies was "promoted with a zeal that blinded its proponents to alternative possibilities and to its inherent shortcomings."[41] His argument is reminiscent of the bureaucratic politics model of foreign policy analysis wherein government departments competing with one another for influence, funds, and power are instrumental in shaping policies best suited to their ends. In this instance, the individual branches of the military compete with one another to legitimize budgets in a zero-sum game, a competition that can produce a "sticky" adherence to strategies that are less flexible to changing circumstances than they otherwise could be.

The first of the strategies Shy identifies, sea power, grew out of the body of thought precipitated by the foundational naval strategist Alfred Thayer Mahan. As Shy writes, the sea power doctrine proscribes that "a concentrated battle fleet could effectively deny the use of the sea to inferior naval forces and could pierce the coastal defense" of an adversary.[42] Sea power was a product of, and co-contributor to, the era of American expansionism and interventionism that swept in a series of small wars and low-intensity conflicts from the Caribbean to the Philippines. It is one of countless examples of how strategy and foreign policy serve to reinforce one another, creating a set of interdependent needs that can, in and of themselves, substantiate the continuation of a strategy. In the case of sea power, the Navy argued that a strong foreign policy required a strong Navy to support offensive capacity, and that a strong Navy required bases abroad (in part because of the twentieth-century need for coal replenishment). As a consequence, American foreign policy goals included entrenching American footholds abroad for the purpose of securing the Navy, for the purpose of an expanding foreign policy. The cycle was self-perpetuating and only ended (to some degree) after contributing to World War I.

None of this is to say that sea power as a governing strategy was necessarily right or wrong. It is to say, though, that even conventional wisdom deserves strong scrutiny, especially insofar as such wisdom conveys institutional benefits to large organizations like the Navy. To this point, Capt. Peter Haynes, USN (Ret.), in his book *Toward a New Maritime Strategy*, lays forth a compelling contemporary portrait of how strategy development in the Navy often remains an exercise in self-preservation dominated by considerations for how the process may impact a service's budget or strategic relevance.

Whether originally a budgetary cudgel, an invaluable strategic perspective, or both, sea power as Mahan explored it is now fundamental to the Navy's identity. Yet today more than ever we see a blurring of the responsibilities of the soldier and the statesmen that threatens the value of stark interpretations of force. The battles fought today are infrequently the symmetric conflicts of warring nations but often the "savage wars of peace" and "dirty little wars" of nation building.[43] In *The Influence of Sea Power upon History*, Mahan asks the reader to ponder the true purpose of a navy. It is a rhetorical exercise, for the answer is obvious; the raison d'être of a navy is the "preponderance over the enemy's ships and fleets and so control of the sea."[44] But what if the enemy has no cruisers or destroyers? What if control of the sea has been achieved, by any meaningful measure, and disorder remains? Such are the questions of maritime security as faced by the United States, which mandate revisiting the conversation on sea power and the very definition of control of the sea.

To be frank, in a book about a criminology-inspired strategy of maritime security, you might reasonably think I could leave big questions of sea power to the experts at the Naval War College, the Naval Academy, the Office of the Chief of Naval Operations, and the Center for Naval Analyses. And to a large extent, after these early pages, I will do just that. Nevertheless, a brief discussion of the broader debate surrounding naval strategy is important when considering the intellectual landscape within which maritime security is situated (or, as is also the case, disregarded). Why? Consider this quote from Gen. Dwight Eisenhower: "If we have the weapons to win a big war we can certainly win a small one."[45] From Vietnam to Iraq, from Libya to Afghanistan, it is obvious that, in this case, Eisenhower was wrong. Navies cannot remain effective against the range of global threats simply by tailoring themselves only for large-scale conflicts. Just as with the weapons and platforms we buy, so too with the strategies we employ. A strategy focused on sea power may not, by default, be useful or practical for maritime security.

We can summarize the major debate in the world of naval strategy as a battle between two camps. One camp believes that the principles of strategy remain unchanged over time. For those who agree with this perspective, "Mahan is (mainly) right, always has been and always will be."[46] The competing camp argues that none but the broadest concepts of strategy extend beyond a given context and that our conceptions of strategy must change as do circumstances.[47] Spurring this divide is the reality that, as a consequence of his prolific writing, Mahan is easily misinterpreted and, as Till notes, "the butt of much unjustified abuse."[48] In truth, Mahan's theories are often more complicated than either side gives his work credit. Nevertheless, the tradition of Mahan lives on in the pursuit of the decisive fleet engagement and the "obvious blue-water preoccupations" of many of the world's largest and growing navies.[49]

And yet Mahan is not the only voice in the traditional naval strategic canon. Julian Corbett's work, for example, remains a popular companion and alternative. Corbett returns Mahan's navy to a broader political and economic context with less of an emphasis on the Mahanian pursuit of decisive engagements and more on the naval role in achieving broader (often joint) military and political ends.[50] Still, both Mahan and Corbett (and their proponents) are frequently read in the lexicon of competition between armadas of conventional adversaries and the accompanying concerns of winning and exploiting control of the sea. No matter how one twists these canonical scholars, we cannot escape the fact that they help us very little in responding to the issues plaguing maritime forces on a daily basis today (human trafficking; narcotics; piracy; illegal, unreported, and unregulated fishing; gunrunning). Hugh White, professor of strategic studies at the Strategic and Defence Studies Centre in Australia (speaking of the Royal Australian Navy), summarizes the point: "Fighting for control of the sea fits Mahan's—and the navy's—ideas about what a navy should be for. Unfortunately it does not fit the reality of naval strategy today."[51]

The conversation, however, does not end with two nineteenth-century theorists, no matter how influential they remain. Geoffrey Till's synopsis of the postmodern navy in his primer on sea power offers a good example of an increasingly popular alternative to these conceptions of sea dominance. Postmodernism suggests that cooperative diplomacy, stability and humanitarian operations, and good order at sea can play instrumental roles in contemporary naval operations.[52] Sea control in this context assumes a function different from what a traditional take on Mahan prescribes. First, the theater of operations for sea control in the postmodern navy is heavily littoral, a terrain with threats patently distinct from the open ocean that Mahanian theorists prefer.[53] Second, postmodern navies conceive of security in a different theoretical framework. Postmodern navies view defense of the global system as a primary objective, what Captain Haynes refers to as a maritime instead of naval perspective. As a consequence, defense against threats to worldwide commerce, shipping, and transportation encourage control of the sea not for manipulation by one state but for the common usage of all licit actors. And the attraction of this vision of sea power has even made inroads in the United States since the fall of the Soviet Union. Former chairman of the Joint Chiefs of Staff (and former chief of naval operations) Adm. Michael Mullen once remarked, "Where the old maritime strategy focused on sea control, the new one must recognize that the economic tide of all nations rises not when the seas are controlled by one but rather when they are made safe and free for all."[54] Thus, postmodern sea power shifts from a focus on dominance to a focus on supervision, opening the door to theories that are better suited to addressing unconventional issues.[55] In the coming years, navies will need to determine how to strike a balance between present and future threats, between modern and postmodern perspectives, "between preparing for a state-centric strategic future as opposed to a system-centric one," and between a conventional high-seas strategy and one of maritime security.[56]

Beyond the macro (institutional and historical) forces that shape the development of Navy strategy and that may serve as obstacles to a greater focus on nontraditional issues like maritime security, it is also important to consider how maritime security missions are disaggregated at a bureaucratic level. An atomized approach to maritime security is enduring because strategists and academics "prefer theories of conflict framed around a single threat—insurgency, terrorism, piracy, narcotics, gangs, organized crime and so on."[57] The outcomes of such an approach, "counter-terrorism, counter-insurgency, counter-piracy . . . might be fine in a binary environment, where one government confronts one threat at a time," or frames it as such.[58] Yet that is not the reality of the complex world we live in. Multidimensional domains demand a theory that is "framed around the common features of all types of threats . . . and considers the environment in toto as a single unified system."[59] Dynamic environments, like littorals or coastal cities, require systemic and threat agnostic approaches to problem solving—a unified theory of maritime security, if you will. As we explore in the coming chapters, broken windows lends a unique and comprehensive framework that fits that prescription.

It is also important to situate ourselves within a broader framework of the existing strategic literature. At the national level, there are several major white papers on strategy, including four prominent ones: the *National Defense Strategy*, the *Quadrennial Defense Review*, the *National Military Strategy*, and the *National Security Strategy* (while my assessment speaks to the documents of the Obama years, recent updates to many of these strategies under the Donald Trump administration have so far only further entrenched the significance of hybrid threats and maritime issues). Throughout this chapter I have so far referenced the 2014 *QDR* and the

2010 *National Security Strategy*. That is not to say, however, that other papers like the National Defense Strategy (*NDS*, 2008) and the *National Military Strategy* (*NMS*, 2011) are not relevant. In both of these publications we see many of the evolving themes noted across this chapter. The *NDS* makes note of the escalatory threat that nonstate actors pose to the international system, the risk of ungoverned or undergoverned territory being expropriated for use by such actors, and the pressures induced by population growth and climate change. The *NDS* even speaks to the instability brought on by transnational organized crime and highlights the need to build capacity across regional partners. The *NMS* includes references to urbanization and population growth in the developing world, the risks to coastal or near-coastal population centers, the growing prominence of nonstate groups, and the role of illicit trafficking networks in aiding the spread of terrorists and weapons. Again, the *NMS* addresses cooperative capacity building, particularly with respect to combating trafficking and smuggling, and even dedicates a short section dedicated to "transnational challenges."

There is, additionally, the *National Strategy for Maritime Security* (*NSMS*), published in 2005. The *NSMS* is a short document (twenty-seven pages, including an appendix). It outlines three principles of maritime security: "preserving the freedom of the seas," "facilitate and defend commerce," and "facilitate the movement of desirable goods and people across our borders, while screening out dangerous people and material."[60] The first and second of these principles are common elements of naval strategy and relate to maritime security only in the very broadest definition of the concept. The last of the three fits squarely within my definition of maritime security but is more an obvious statement of homeland defense than international strategy. In fact, while the *NSMS* does mention transnational crime and piracy, its primary international focus is on the threat of terrorism. Most of the document is informed by the four objectives derived from these three principles: prevent terrorist and criminal acts, protect critical infrastructure and population centers, mitigate risks, and protect the marine environment. And yet the result is not so much an international strategy but a domestic threat and vulnerability assessment. Shoring up critical infrastructure, reducing the risk posed to significant targets, reducing recovery time in the event of an attack, and denying an adversary the opportunity to recognize and strike valuable assets—all found in the document—are components of a threat and vulnerability assessment. The *NSMS*, therefore, is an example of how maritime security could be addressed on a technical and strategic level within the United States and across ports servicing the United States. But, while it broaches the importance of layered security and combating unconventional threats (most notably terrorism), the strategy predominantly speaks to global shipping and drifts more toward aspirational statements in the broader international security context. The *NSMS*, therefore, brings us no closer to a serious discussion on a comprehensive strategy of littoral security.

At the force level, I delve deeper, below, into one capstone maritime strategic document, *A Cooperative Strategy for 21st Century Seapower* (*CS21*).[61] In addition, there exist still other important naval and maritime publications and statements, including *Naval Doctrinal Publication 1* and *Naval Operations Concept 2010* (*NOC*). While maritime security is not the focus of *Naval Doctrinal Publication 1*, there is still a short section on the need to build international capacity to combat transnational traffickers and terrorists using the seas for illicit purposes. The *NOC*, however, includes a whole chapter on the subject, which discuses many of the issues explored above. The *NOC* was intended to give substance to *CS21*'s broader strategic statements. It is therefore noteworthy that even here we see a lack of coherent strategy with respect to maritime security. Responses to seaborne threats continue to be framed as "counter-something." Counterpiracy, counterterrorism, drug interdiction, migrant interdiction, and more become the cornerstones of a half dozen unique and distinct international maritime security missions. Thus, even where maritime security receives long overdue attention, the way in which it is framed remains tied to combating individual acts. This is more doctrine than strategy. And to borrow an analogy from Till, "if maritime theories are about the art of cookery, doctrine is concerned with today's menu."[62] Put another way, strategy is how a military chooses to prioritize perspectives and leverage military might to meet larger political objectives, while doctrine is more concerned with connecting this lofty conjecture to the practical reality of operating day to day. Both are important, but without the wider strategic framing, doctrine is inevitably shortsighted.

Interestingly, this framing problem in the strategic literature is mirrored in the structure of security studies. Subfields on terrorism and piracy have grown rapidly, offering immeasurable value in their detailed investigations of specialized topics. And yet the growth of subdisciplines has also led to the ghettoization of literature, and we run the risk of increasingly ignoring the gray space in between as a matter of disciplinary purity.

A compounding, contributing factor to the fragmentation of literature on maritime security, piracy, and terrorism is their relative newness.

Maritime security, in particular, has risen in prominence alongside a renewed, post-9/11 interest in nonstate threats. As Basil Germond explores, prior to 2002, the term "maritime security" was virtually nonexistent in academia. Of the more than 16,000 Google Scholar hits for the phrase "maritime security," more than 200 were penned between 1914 and 1988.[63] Since 2002, the term's usage has increased meteorically. We can also see this emergence in the strategic literature. Later I discuss, in the American domestic context, the role of *CS21* in raising the prominence of maritime security for the Navy, Coast Guard, and Marine Corps. In the international context, strategic documents have been even slower to emerge. NATO was the first of the major intergovernmental organizations to publish a modern maritime strategy document, *Alliance Maritime Strategy*, in March 2011. While the document hits all the right notes, mentioning piracy, terrorism, weapons of mass destruction proliferation, transnational organized crime, and even global warming, it does so in a notably abbreviated fashion. NATO does not define what it means by the "maritime environment" or "maritime security," which, alongside the strategy's brevity, may suggest that member countries failed to reach a deeper agreement on what such terms signify. One year later, the African Union set out its own agenda in *2050 Africa's Integrated Maritime Strategy*. Far longer than the NATO document, the African Union's strategy is focused on rectifying the continent's ongoing struggle with establishing maritime domain awareness. Its authorship appears motivated largely and logically by the untapped wealth of Africa's long coastline, yet at times the document bridges into the obscure (nestled halfway through is a section titled "Giant Africa Aquariums"). Yet, more than anything else, the dominant impression from this comparatively lengthy text is that it is more a wish list than a strategy. The African Union acknowledges the continent's maritime porosity but requires too much of a fundamental investment in capacity building to implement any comprehensive strategy.

Finally, the European Union published the *European Union Maritime Security Strategy* in June 2014. It is a relatively comprehensive document, strongest in its attention to the technical details of integrating so many member states. While the strategy succeeds in presenting many employable proposals, the ongoing EU tension between interoperability and sovereignty remains an undercurrent throughout the text. Yet what is significant about these documents is not so much their content but their dates of release. Maritime security is clearly a new and evolving field. Despite the fact that the illegal trafficking of people is an industry worth billions of dollars, despite the fact that a huge portion of interdicted drugs are seized in the maritime environment, and despite the fact that some terrorist organizations use the sale of narcotics or other smuggled products to fund their campaigns, maritime security has only recently emerged on many countries' radars as a strategic problem. Only in the last thirty years has the international community come together to address maritime drug smuggling, and the response to human trafficking was tragically slow.[64] Of the roughly one dozen international treaties on terrorism signed in the decades before September 11, only one (the Convention for the Suppression of Unlawful Acts against the Safety of Maritime Navigation) addresses acts of terror at sea.[65] For those willing to wade into the muddy waters, maritime security offers the opportunity to substantively contribute to an evolving field—there is a lot of work yet to be done.

Strategy versus Acquisition

While some publications, including the *QDR* and the *National Security Strategy*, have begun addressing maritime security in broad strokes, these efforts often come in fits and starts. In the 2009 joint publication on amphibious operations, the words "urban" and "city" are absent, and mention of littoral urbanization is absent in a 2012 doctrine on joint forcible entry.[66] For more than a decade, the American military has focused on a very narrow type of counterinsurgent and counterterror conflict. Yet, slowly, the military has come to understand the importance of transitioning to a broader understanding of security. As the United States struggles to extricate its forces from the Middle East and Southwest Asia, strategists are likewise extricating their minds from the deserts and mountains (to borrow a phrase from Kilcullen) to resume a conversation on littorals that was emerging before 9/11.[67]

After the fall of the Soviet Union, the Department of the Navy published . . . *From the Sea* and *Forward . . . From the Sea*. These publications wrestle with the responsibilities of a virtually uncontested Navy in a unipolar world and highlight a transition from securing exclusive control of the sea (largely a given by then) to securing maneuver space in regional littorals. Their subsequent guidance aims to ease the inevitable transition "as Naval Forces shift from a Cold War, open-ocean, blue water naval strategy to a regional, littoral, and expeditionary focus."[68] For a more detailed assessment of this transition, Peter Haynes chronicles the development of these themes deftly, and the preponderance of publications on Navy capstone strategies by Capt. Peter Swartz, USN (Ret.), is a wonderful resource. In brief, however, the Navy in the 1990s was preparing to retool and tackle the threats not of an imposing Soviet armada but of the complicated geostrategic role of regional littorals (albeit still oriented against national threats). The attacks of 9/11 interrupted that transition, but the maritime services began to resume that theme in the 2010s. In 2011, the *Marine Corps Operations* manual emphasized "complex expeditionary operations in the urban littorals" and the "shifting strategic focus to the littoral regions."[69] In 2012 the Navy created the Coastal Riverine Force to help combat threats particular to brown waters.[70] And in 2007 the maritime services (the Navy, Coast Guard, and Marine Corps) published *A Cooperative Strategy for 21st Century Seapower* (later to be revised in 2015), which demonstrated one of the most significant steps in the evolution of the strategic importance of maritime security. To analyze the implications of this evolution, it is useful to juxtapose *CS21* against the Navy's approach to acquisitions. In other words, when talking about maritime security, did the Navy put its money where its mouth is?

CS21 is the first joint publication of its kind. It attempted to connect the sea services under an umbrella strategy, which is particularly noteworthy for its allusions to Admiral Mullen's articulation of the postmodern naval principle of defending the global system.[71] The strategy is replete with allusions to a wider view of maritime strategy, inclusive of the role of stability and constabulary responsibilities. This includes a rejection of the binary perspective on war and peace in favor of an understanding that preventing conflicts is as important as winning them. (This latter point, a commitment not just to war fighting but to war prevention, is evocative of a related debate in policing over crime fighting and crime prevention, to which I return in the following chapter.) The document even embraces discussion of many of the unconventional themes and issues discussed above. Thus, while the focus of the U.S. Navy on deterring and, if need be, winning large-scale conflicts was never in doubt, the strategy also wrestled with the growing significance of crowded, networked, and impoverished littorals. There are, consequently, some underlying tensions in the strategy, which at times seems to speak almost of two strategies divided along high-end and low-end types of conflict. Haynes, in his discussion on how the strategy came to exist, underscores this tension between postmodern and traditional impulses among the text's authors, which may contribute to the sense that two divergent strategic visions inhabit the same document. Still, *CS21* showed that the Navy is willing to consider that two distinct geopolitical challenges—defense of the global system against peer competitors and maintenance of the global system in light of unconventional threats—might best be met by two distinct approaches to strategy, even if only by virtue of the deliberative process that created the final publication.

The capstone document divides these two approaches into "Regionally Concentrated, Credible Combat Power" and "Globally Distributed, Mission-Tailored Maritime Forces." Each is then further subdivided into several functional obligations. For the former group, these responsibilities include containing regional conflict through forward deployed forces, deterring war among major powers, and winning wars when necessary. This first approach frames the Navy's traditional self-perception. It is for this approach that the newest *Gerald Ford*–class carrier was constructed, and likewise it is for this mission set that major acquisition projects like *Columbia*-class submarines or *Zumwalt*-class destroyers were designed. Admittedly, maintaining the capacity to win large wars will inevitably require expensive hardware. Still, keeping in mind the dictum that optimizing for big wars does not guarantee winning small ones, it is important to frame these acquisitions in their opportunity cost to maritime security. That does not make them bad (or good) decisions, but part of bringing maritime security to the strategic table means ensuring that those costs are acknowledged. Thus, while the text of *CS21* is equally split between the traditional and postmodern, we can see already that there remains a considerable institutional emphasis on the traditional. That institutional preference has been further dramatized in the years since *CS21* was revised under CNO Jonathan Greenert and then replaced by CNO John Richardson's capstone document (*A Design for Maintaining Maritime Superiority*). In that time, the prominence of maritime security has backslid to make room for a heightened focus on China and Russia.

Back to *CS21*, the document's second approach—globally dispersed, mission-specific forces—was a demonstrative conceptual step toward strategically exploring maritime security. Here *CS21* emphasizes the need to respond to natural disasters, climate change, nonstate actors, social instability, criminal elements, and forced migration. As before, this category is subdivided into several functional obligations, including defending the homeland, cooperating with international partners (on operations such as humanitarian assistance and law enforcement), and combating local disruptions (threats such as piracy, human trafficking, drug smuggling, and arms trafficking) before they evolve into regional or global crises. These three responsibilities underscore the awareness that "increasingly urbanized littoral regions" and the unconventional threats they engender are fast becoming major points of concern for actors a world away.[72] In this arena as well, Navy acquisitions paint an insightful picture. The littoral combat ship (LCS), with its shallow draft, comparatively expansive hanger space, large flight deck, small crew, fast speed, and sophisticated automation, was designed to become the low-end workhorse of the fleet. Even over budget, the LCS ostensibly provides a cost-effective alternative to sending a multibillion-dollar destroyer into littorals to combat pirates, terrorists, or traffickers in their dinghies and open-hulled speedboats. The LCS's small size optimizes its capacity to coordinate with other navies and coast guards, many of which do not have the port facilities, ships, or personnel sufficiently skilled to train with mammoth vessels. The LCS's shallow draft also allows it to enter hundreds more ports than a larger ship, and some analysts contend that its speed and integration with manned and unmanned airpower increase the ship's resilience to asymmetric threats.[73]

Yet the LCS has been repeatedly targeted for cancellation or truncated production. In an environment of budget austerity, resources devoted to *CS21*'s second approach were continuously seen in opportunity cost of ships designed for combat with a conventional adversary in mind. Truly, much of the debate surrounding the LCS stemmed from legitimate criticism of the ship's performance and the Navy's apparent build now, strategize later philosophy.[74] And yet the intensity with which the program was targeted may exceed the standard criticism that routinely accompanies the launch of new classes of surface combatants. Was there something different about the debate surrounding the LCS? While the ship design may have required modifications, critiques of its performance may also augur a deeper criticism, reflective of clashing schools of thought between the proverbial "big Navy" and its postmodern wing. This criticism represents in part a failure to express or embrace the duality of the Navy's responsibilities, as first detailed in *CS21*. For example, it is largely uncontroversial to assert that the LCS would be on the losing end of an engagement with the Chinese navy, as former secretary of defense Chuck Hagel once noted.[75] The LCS, however, was not (or should not have been) built to stand against the PLA Navy. The Navy is in possession of a substantively and theoretically distinct ship than the classes of warships it is set to replace, and that is "precisely what the Navy wanted."[76] That the office of the secretary of defense saw the ship in a dramatically different light suggests the long-standing difficulty on the part of the Navy's postmodern wing to effectively articulate (and build support for) the differences between its two missions of defending the global system and maintaining it. The LCS is optimized for an entirely different set of missions than engaging the Chinese navy, and its effective integration into the fleet rests on "a division of labor . . . in which different classes of warship are configured for different levels of danger."[77] The ship could, therefore, play a significant role in the second of *CS21*'s two missions, that of international maritime security. To this end, even as the exact status of the LCS's final design was subject to rigorous debate, the Navy's commitment to supporting some LCS variant may signal, through an optimistic interpretation, that the fleet sees the need to foster new ways of cost-effectively performing maritime security missions.

Ultimately, and thankfully, neither maritime security operations nor the articulation of a maritime security strategy rest on the fate of the LCS. Larger ships have their own advantages in humanitarian assistance and disaster relief and related missions, not to mention that any strategy of maritime security would be impossible without the command of the sea that larger platforms and capabilities provide.[78] Still, the narrative around the LCS may offer a window into deeper strategic questions regarding how, if at all, the Navy reconciles budgetary priorities with the conceptual call heralded by *CS21* to not only defend but maintain the global maritime order.

For the time being, "most of the warships in the present and planned fleet aren't well suited for littoral operations," either because of lack of speed, cost of operations, or draft.[79] As a consequence, when read through the lens of recent acquisitions, we are forced to a more tepid conclusion about the long-term significance of *CS21* and what it meant for maritime security strategy. The Navy's continuation of LCS production (in a modified form), the continued prominence of maritime security in the revised version of *CS21* (though far less so in Chief of Naval Operations Richardson's *Design*), and an ongoing robust debate on transnational issues, all suggest that the strategy was more than just lip service to the growing importance of littorals and unconventional threats. And yet programmatic realities make it obvious that *CS21* falls far short of making these issues a driving priority. Former secretary of the navy Donald Winter was in fact emphatic, and far from alone, in his assertion that *CS21* would in no way influence fleet architecture, and no associated force structure plan was ever developed to compliment the strategy.[80] (This isn't news to strategists, who struggle to inform, let alone drive, the Navy's composition.) Still, even as associated investment lags, there can be no doubt that strategists are increasingly talking about the risks posed by the unconventional. As John Nagl notes, large conservative institutions require time and concerted effort to change and innovate.[81] That the Navy is even discussing hybrid threats, global warming, and the illegal trafficking of people, drugs, money, and weapons is an encouraging sign for maritime security strategists.

❖ ❖ ❖

So, why does maritime security deserve the attention of the naval strategy community? Why should the U.S. Navy, in partnership with its sister maritime services, take maritime security seriously as a body for strategic development and investment in its own right? The increasingly hybridized nature of urban coastal threats will only compound preexisting stressors like climate change and poverty in the coming decades. Demographic and human geographic changes are placing new and significant stress on the maritime domain, and projections indicate that these factors will continue to drastically reshape international security. The resulting pressures those forces will bring to bear on the U.S. Navy are unlike the traditional conflicts Mahanian strategies are optimized to address. As a consequence, this has also begun to drive small but important change in proverbial big Navy's long-standing ambivalence toward these issues. While not groundbreaking in programmatic terms, *CS21* was an invitation for "further experimentation," while documents like the 2014 *QDR* spoke directly toward the growing influence of nonstate actors in an interconnected world.[82] It is the need to engage with this experimentation that makes maritime security relevant. The trends introduced above—urbanization, poverty, globalization, global warming, and development in the littorals—will become only more prominent throughout the coming decades. While the Asia-Pacific may continue to dominate the Navy's attention and resources in the years ahead, developing an effective strategy of maritime security will increasingly define how the Navy performs globally.

In a domain where "the international architecture of the twentieth century is buckling under the weight of new threats," the risk of disorder and poor governance is among the rising themes in American national security publications, should they not be drowned out by the specter of specific national adversaries.[83] As chronicled in strategies from the White House, Department of Defense, and the maritime services, ensuring law and order in the face of threats that recognize no boundaries is among the leading problems of the twenty-first century. Developing a strategy of maritime security capable of addressing the instability wrought by these threats requires new ways of thinking about what the Navy does, and where.

In the introduction to this book, I finished with the most fundamental question for any strategic initiative: why should we care? By way of response, in this chapter we looked at demographic and human geographic trends, gaining a better understanding of the global threat forecast over the coming decades; at how the Pentagon is thinking about the future; and at how the U.S. Navy is writing about (if not always acting on) some of these issues. Throughout, we saw hints of how policing might intersect with hybrid threats, and how broken windows' contextual and communal approach to crime prevention could prove a useful foundation for a strategy of maritime security. It is to this topic, community policing and the broken windows theory, that I turn to next.

New York Police Department vehicles line Pier 88 in Manhattan as the amphibious assault ship USS *Kearsarge* (LHD 3) begins its departure, signaling the end of 2017 Fleet Week New York. An event spanning more than two decades, Fleet Week New York is the city's time-honored celebration of the sea services. It is an unparalleled opportunity for the citizens of New York and the surrounding tristate area to meet sailors, Marines, and Coast Guardsmen as well as witness firsthand the latest capabilities of today's maritime services. *U.S. Navy photo by Mass Communication Specialist 2nd Class Carla Giglio/Released*

CHAPTER TWO
Breaking Windows

THE POWER OF CONTEXT

At the center of the broken windows theory is the notion that combating crime does not necessitate the immediate resolution of root causes such as poverty or substance abuse. By changing people's environments, both physical and perceived, through addressing small-scale disorder, even large-scale epidemics of crime can be curbed. On the New York City subway, cracking down on fare beating resulted in a dramatic turnaround for what was once one of the country's most violent transportation systems. Above ground, policing quality-of-life misdemeanor crimes such as graffiti or aggressive panhandling helped reduce felony rates to their lowest level in years. In the maritime environment, understanding issues through the broken windows lens means placing a greater emphasis on understanding the interplay of perception and security. Doing so may look like offering more humanitarian assistance, combating human trafficking, or targeting narcotics and arms trafficking. While these are catastrophic on a human scale, constabulary duties have always been small fish for big Navy. Yet, as we will see in this chapter, broken windows theory suggests that dealing with issues on the lower end of the threat spectrum can have reverberating impacts across that spectrum. Nowhere is this simplicity of disproportionate impact examined more articulately than in Malcolm Gladwell's *The Tipping Point: How Little Things Can Make a Big Difference*. In it Gladwell speaks of epidemics, which might include everything from fashion trends to syphilis outbreaks, and identifies three factors that drive them: the law of the few, the stickiness factor, and the power of context.[1] One factor in particular played a prominent role in the crime epidemic sweeping U.S. cities in the late 1980s and in bringing the crisis to heel: context.

The power of context, at its most basic, stipulates that our environment influences all of us. That much seems obvious; clearly context impacts our lives. What is unexpected is just how profoundly and subtly environment and perception shape the decisions we make. Clinical and psychological studies on everyone from preteens to seminarians suggest that our surroundings have an almost overriding impact on how we behave. Students may cheat on math tests but not spelling tests, or they may cheat at home but not at school, but there is nearly always a context in which a student will cheat.[2] Their environment informs how and when students break the rules, but it is almost inevitable that, in some combination of circumstances, any one of us would cheat on an exam. Likewise, in an experiment, seminary students who were told they were late for delivering a sermon were more prone to literally stepping over a man in need of help (an actor) than ones told they were early.[3] Ironically, both groups were told the subject of the sermon should be on the parable of the Good Samaritan. Context was the deciding factor here as well. Seminarians literally stepped over people in need while running to deliver a sermon on the Good Samaritan simply because they *felt* rushed. Subtleties of context, even if that context is purely psychological, do not merely inform our actions, they often instruct them. The broken windows theory and Gladwell's power of context are synonymous. By shaping an environment, broken windows theorists speculated that they could tip favorably, and maybe even reverse, epidemics of crime. Context can be changed and, therefore, so can actions—or so the theory holds.

Theory and Terms
The *Atlantic* magazine is more than 160 years old. During its tenure as an opinion shaper in American history, the magazine has published everyone from Mark Twain to Martin Luther King Jr. It is therefore no small testament to George Kelling and James Wilson that, among these figures, it is the piece they published in March 1982 entitled "Broken Windows: The Police and Neighborhood Safety" that endures as one of the most reproduced articles in the publication's history.[4] The forceful simplicity of their argument and the sensitive nature of police policy in the turbulent 1980s combined to catapult broken windows into a major competitor to traditional views of police work.[5] While Kelling in particular would go on to publish more specific elucidations of broken windows—notably with Catherine Coles in *Fixing Broken Windows*—this first article seemed to strike a chord in and beyond the field of criminology.

That chord resonated in part because the foundation had been laid thirteen years earlier with a transcontinental experiment.[6] In 1969 two cars were parked and abandoned with their hoods up. One was in Palo Alto, a wealthy California suburb; the other was left in the Bronx, a rough borough of New York City. Within ten minutes, individuals had set upon the car in the Bronx. Within twenty-four hours, there was almost nothing left of it to take. The car quickly became a jungle gym for children, just a piece of rust on a dirty Bronx street. The fate of the Bronx car seems fairly obvious to anyone with a passing familiarity with what was one of New York's grittiest neighborhoods. The Palo Alto car fared much better for a while, until Philip Zimbardo, the Stanford psychologist responsible for the experiment, hit it with a sledgehammer. Soon the good-natured people of the Golden State had destroyed the vehicle and left it entirely overturned.

The varied mindsets of Bronx residents—ranging from indifferent, to discouraged, to utter disregard for property—facilitated the early onset of vandalism there; the people of the neighborhood expressly believed no one cared what happened to abandoned cars. In Palo Alto, the community expectations were different. Law and order were well maintained, and people had dramatically different experiences and expectations of behavior. Yet, even in that comfortable suburban enclave, the smallest signal that no one cared (i.e., the sledgehammer to the car) dissolved communal restraints almost instantaneously.[7] The experiment, while preceding the *Atlantic* article, is a proof of concept for broken windows. It tells us not just that perceived environmental decay encourages the rapid disintegration of community self-policing but also that no neighborhood is immune to the power of context.

The message, however, is not fatalistic. If Zimbardo had not smashed the car a week into the experiment, who knows how long Palo Alto's residents would have continued to walk past it. Instead, the intent is to uncover why some communities precipitate behavior that results in action such as that in the Bronx and how they may be altered to mirror Palo Alto. Criminologists have posited several theories about socioeconomics, narcotics, alcohol, and more to explain the vicissitudes of certain neighborhoods. While many highlight serious social issues that deserve attention, the broken windows theory posits that none of these root-cause explanations shares the same level of correlation with a community's crime rate as does the prevalence of disorder.

Addressing disorder, often referred to as broken windows policing or quality-of-life policing, is the practical manifestation of the relationship between environment and behavior. As cases in the New York subway and New York City highlight, law enforcement's ability to leverage the hypothesized association between context and action produced a measurable result on the rate of violence.

But how does one dent in an abandoned car or one derelict building truly spark disorder, even crime? To understand this relationship more clearly, let's take a more narrative approach that traces the process step by step. Wilson and Kelling sketch the following example: "A piece of property is abandoned, weeds grow up, a window is smashed. Adults stop scolding rowdy children; the children, emboldened, become more rowdy. Families move out, unattached adults move in. Teenagers gather in front of the corner store. The merchant asks them to move; they refuse. Fights occur. Litter accumulates. People start drinking in front of the grocery; in time, an inebriate slumps to the sidewalk and is allowed to sleep it off. Pedestrians are approached by panhandlers."[8] That almost sounds like Archie Bunker's worst nightmare, and equally as dramatic. The truth is, however, that this outline has repeatedly proven illustrative of real-world cases. Disorder spreads steadily. First, a window is broken. If it goes unfixed, soon all the windows of that building will likely be broken.[9] Why? The destruction of property surely does not aggregate in certain districts because there is a disproportionate distribution of window-smashers to window-lovers; rather, the first broken window serves as a signal, like a sledgehammer to a car.[10] It is a sign that no one cares, no one is watching, and there is nothing to stop someone from breaking the next one. Kelling and Wilson's aggressive panhandling is also a broken window. As they note, muggers prefer to operate where people are already intimidated; they believe the chances of detection are lower where residents cannot even seem to keep their streets orderly. They begin to consume the area; eventually a mugging goes wrong and the neighborhood experiences a murder. Suddenly a series of events that started with an abandoned property ends in violent crime. The simplicity of the metaphor of a broken window belies the complex mechanisms of a community's ability to self-monitor and law enforcement's ability to impose order, which a single signal of minor vandalism can in time unravel.

Why was broken windows theory needed? Nationwide crime rates for the 1970s through the late 1990s certainly suggest something was not working. Kelling and Coles make the argument throughout *Fixing Broken Windows* that law enforcement's midcentury crime-fighting approach was an ineffective way of addressing the damage caused by disorder. Such a style of policing is likely familiar to the reader. It is typified by police in their cruisers, always available for dispatch calls, reluctant to engage with pedestrians, and directly attached to computerized 911 networks.[11] Police and the communities they serve grow apart, distrust builds, and disorder is compounded by inaction and a sense of loyalty to the profession.[12] For example, many felt that the New York Police Department's (NYPD) guiding philosophy during this time was plainly, "stay out of trouble."[13] Kelling and Wilson's suggestion was to control crime through controlling disorder, not trying to catch every crime in the act. Order maintenance (quality-of-life policing) is the effort to constrain the behavioral and physical manifestations of disorder, which serve as signs of neighborhood decay.[14] This is crime prevention, not crime fighting. By stemming disorderly conduct and perceptions thereof, the process of degradation can be intercepted, and order maintenance can break the cycle. Most empirical research studying broken windows policing uses misdemeanor arrests as a measure of quality-of-life policing. Such crimes include typical low-level malfeasance such as graffiti, vandalism, public drinking, and so on. Importantly, Kelling's definition of broken windows policing accounts for the influence of more than just the physical environment on our actions.[15] It is inclusive of a holistic understanding of context, to include behavior alongside more literal broken windows, as is Gladwell's power of context. It is the intention of this research to apply that same perspective to maritime security by assessing how the physical and behavioral environment of littoral communities intersects with crime and instability.

Notwithstanding the obvious need for innovation in the police force, this alternative community-based policing model was greeted with strong criticism. Kelling and Coles recount how police simply could not be bothered to address minor issues when there was "real" police work to be done. A half century of reform in police strategy had led many to believe there had never been any other function for law enforcement besides crime fighting. This cognitive dissonance is not uncommon in large institutions with their own cultures and agendas. In the military, for example, "a remembered past has always more or less constricted both action in the present and thinking about the future," regardless of the accuracy of that remembered past.[16] In reality, before the mid-twentieth century, police were seen as primarily responsible for "crime prevention" and order maintenance, "keeping peace by peaceful means."[17] Even the very phrase "law enforcement" was not in use until the 1960s, around the time that the idea of the criminal justice "system" first found expression.[18] As a result, police were thrust into a systematized bureaucracy that effectively reduced their role to the apprehension of suspected criminals.

Another significant point of friction was the legal pushback from advocacy groups that see order maintenance tactics as veiled attempts to target the rights of homeless individuals.[19] Kelling and Wilson were frequent targets of attacks of classism and racism, prompting Kelling and Coles to directly address the subject of homelessness in their book, *Fixing Broken Windows*. By presenting what they felt was a nuanced understanding of the homeless community, Kelling and Coles hoped to substantiate their focus on order maintenance while demonstrating that their philosophy was ultimately a positive for all city residents, including those most vulnerable. While the successful implementation of community policing programs quickly won over officers, legal challenges threatened to make these gains obsolete. This was perceived as such a threat to the survival of broken windows policing that Kelling and Coles dedicate a full chapter, and broad sections of many others, to discussing the legal support for their position and explaining some misrepresentations they deem typical of suits filed by advocacy groups on behalf of the homeless. The compelling legal arguments on both sides have resulted in pitched court battles in nearly every city or state that has implemented order maintenance legislation (Seattle, San Francisco, New York, Texas, Baltimore, Miami, etc.).[20]

Finally, community policing undeniably sounds "soft," a tone belying order maintenance's proactive approach to crime prevention.[21] Some law enforcement officials consequently regarded broken windows as beneath their profession or beyond the scope of their responsibilities within the criminal justice system. This too is paralleled in the armed forces. Order maintenance is relegated to "operations other than war," at best something to do while waiting for a real war, at worst a distraction and a waste of resources. Yet order maintenance reached receptive minds among police. New York City police commissioner William Bratton, serving in that post from 1994 to 1996 and again from 2014 to 2016, quickly became a broken windows adherent. Bratton would prove critical in demonstrating on a massive scale the potential of order maintenance policing, uncovering associations between disorder and violent crime that Kelling and Coles were only just beginning to notice. Two cases chronicling Bratton's stewardship of New York City's Transit Authority Police Department (TPD), and later the NYPD, illustrate the power of Wilson and Kelling's concept when put into practice. These cases are also two common studies in broken windows literature, and therefore an understanding of each covers a useful scope of the field. While not the first implementations of order maintenance, New York's experimentation was among the largest and most potent demonstration of an early foray into broken windows.

Before moving further with historical cases though, it's worth taking a moment to place broken windows in a more contemporary context. In 2014 a wave of police shootings in the United States in which mostly white officers killed or injured unarmed black men precipitated protests throughout the country. In New York City demonstrators demanded an end to broken windows policing, charging that the theory is both racist and criminalizes the homeless (the latter issue extensively tackled by Kelling and Coles). And in practice, particularly in the use of stop and frisk (which, I would argue, is anathema to the aims of community policing and the broken windows theory), the institution of policing across the country has demonstrated a worrisome trend of racial bias. The popular dissatisfaction with the broken windows theory is understandable given the association New Yorkers in particular have between the term and allegations of police misconduct. Nevertheless, in this body of work, I'll maintain a distinction between the more colloquial use of "broken windows," popularized in 2014 as shorthand for police impropriety, and the theory as developed by Kelling, Wilson, and Coles and implemented under Kelling's direction in the late 1980s and early 1990s. More on the broken windows in its contemporary incarnation can be found in the conclusion.

The New York City Subway

A day in the life of an average New York subway user in 1984 was trying in the best of times. Trains were consistently late due to daily fires, trains derailed twice a month, train cars were covered in trash and graffiti, and efforts to dodge transit fairs were estimated to cost the Transit Authority up to $150 million a year.[22] On any given day the system experienced more than forty felonies (it would top fifty a day by 1989), and aggressive panhandling and the pervasiveness of petty crime had forced ridership to its lowest point in history.[23] Extortion of passengers was common, with vandals disabling token machines and collecting fares from customers as they held open gates to the platform. In fact, the chairman of the New York City Metropolitan Transit Authority (MTA), Robert Kiley, and the future TPD commissioner, Bratton, were both caught in traps such as this.[24] In short, life on the subway was dirty and dangerous. Then suddenly, from a record-high felony rate, subway ridership improved and violent crime dropped 75 percent.[25] How?

In the 1980s, Kelling began consulting for the MTA. Around that time (1984) the newly appointed Transit Authority president, David Gunn, launched the Clean Car Program, a multibillion-dollar initiative to eradicate graffiti from the subway. As broken windows theory suggests, graffiti on a train provides "inescapable [proof] that the environment [a rider] must endure for an hour or more a day is uncontrolled and uncontrollable, and that anyone can invade it to do whatever damage and mischief the mind suggests."[26] It took military diligence to take back the more than five thousand cars used by 3.5 million people a day. Special police units rode in and defended reclaimed trains, and any clean car with new graffiti was either fixed in the short changeover at the terminus of a route or removed from the tracks.[27] It was not until May 12, 1989, that the final subway car was cleaned, at which time the MTA had accomplished one of the largest urban restoration projects ever attempted.[28] And yet crime persisted. Was broken windows broken?

Not exactly. While the graffiti removal was prerequisite to taking back the subway, it was not the end of the line. Disorder had become so rampant in the system that even this massive undertaking was not enough to bring it under police control. The deeper problem had always been broadly labeled homelessness, "and homelessness was not the TPD's problem."[29] It was not until Kelling headed an investigative study for the MTA that the issue of disorderly criminal behavior was attributed as the real issue.[30] Once that occurred, officers were given the tools to begin tackling the appalling conditions of their underground world. In October 1989 Operation Enforcement kicked off to a slow start, and in April 1990 William Bratton took the reins of the TPD.[31] Like Gunn, Bratton was a broken windows disciple, and his leadership saw an instantaneous spike in enforcement; "ejections for misbehavior tripled within a matter of months after Bratton took office."[32] The primary target? Fare beaters.

In 1990 a quarter of a million people were avoiding paying fares *every day*, and, like graffiti, Bratton saw fare evasion as a signal of disorder.[33] Despite costing millions of dollars to the city annually, TPD officers had been reluctant to pursue those hopping turnstiles. It took almost an entire day to process someone for a $1.25 misdemeanor; it just wasn't worth it. Bratton sought to change that. He remodeled a transit bus and turned it into a mobile station house. Officers had fare beaters stand in daisy chains until they had a full haul; soon processing times were down to an hour even as arrests increased, and the tide began to change. The full impact of order maintenance became immediately apparent; one out of every seven fare beaters arrested had an outstanding warrant, and one out of every twenty was carrying a weapon.[34]

Put another way, police found direct expression that crime is multidimensional. About one in ten arrests for a $1.25 offense yielded a felony or Class A misdemeanor arrest.[35] Nothing could have better exposed the links between disorder and crime than the proof that many fare beaters were not just misbehaving delinquents but possible felons or career criminals. Crime drops in direct correlation to an increased effort to maintain order because such policies put police in contact with violent offenders before any more offenses are committed.[36] Crime, and the individuals who engage in crime, is multidimensional. To their credit, "the bad guys wised up," weapons were left at home, and everyone started paying their fares.[37] By the early 1990s crime was no longer a serious problem on the New York City subway, and before the decade was out, the system could be said to be "among the safest in the world."[38]

Some may argue that the subway case is too narrow an example to validate broken windows or use it to apply the theory to the littorals. Although New York's subway is more extensive than most, it is in the end a "spatially bounded" network with defined points of ingress and egress.[39]

That, however, is also an advantage, serving as a contained natural experiment isolating many of the factors that are difficult to control in a city. The reclamation of the New York City subway was one of the largest municipal problem-solving exercises ever attempted, and its spatial restrictions make Bratton's success difficult to attribute to anything other than direct police policy. Even more so, the root cause explanations often attributed to the decline in citywide (and nationwide) crime later in the decade do not apply: the subway was not a major drug market, the economy was still struggling, unemployment was on the rise, and the youth population had not shrunk. Target hardening, access restriction, and managerial changes all played a role. But it was not until order maintenance was properly enforced that a "tipping point" was reached and there were any noticeable changes in subway crime. The closed system gave broken windows an opportunity in which the many distracting variables endemic to large cities were naturally controlled.[40] As a consequence, the physical environment was restored and the runaway behavior was curbed.

Among the most significant problems with taking broken windows into the maritime domain is the transition from urban to open. Even studies that have adopted broken windows in alternative disciplines have applied the theory overwhelmingly in urban settings. Wilson, Kelling, William Sousa, and Coles have produced penetrating insight into the machinery of order maintenance, but their work remains tied to the metropolitan unit. No two cities are identical, but the existence of neighborhoods and communities is universal, even if their borders are ambiguous. The New York City subway experiment, however, suggests that habitation is not a prerequisite for the application of the broken windows theory. Just as people do not generally live on the subway but rather use it as a means of transportation or employment, so too do people rarely live on open water but instead use it as a means of transport and economy. The mere presence of people in an area creates its own community with its own rules, or lack thereof. Like our "streets, parks, and sidewalks" the subway and waterways "belong to no one and therefore to everyone."[41] Moreover, over the past several decades, the spheres of criminology and international security have been in slow convergence, as seen in the previous chapter. As the world's growing population continues to crowd ever more people into sprawling slums and shantytowns, urban issues such as poverty and crime have slowly become crises of international proportions. The ongoing worldwide movement toward urbanization and littoralization—the tendency for people to cluster along the water—makes it clear that theories of security will increasingly need to borrow from theories of urbanization.

The subway is one very small example relative to the global scope of this book. Nevertheless, its recovery lends strong support to broken windows as a theory of an overall behavioral, not strictly urban, condition. Broken windows theory illustrates the mechanics not only of urban decline but of human psychology. Indeed, the consensus among sociologists and police that one broken window inevitably leads to more is not dependent on the urban political-geographical setting of window breakers but rather on their immediate context. This is Gladwell's approach as well; *The Tipping Point* and the power of context are about creating, stopping, and altering epidemics on scales that are unencumbered by the confines of points on a map. In expanding broken windows beyond the basic city unit, I may run the risk of overextending its language; however, the literature supports the position that there is no concomitant risk of diluting the relevance of its principles.

From the City to the Sea
Broken windows is a problem-oriented theory. Its principles are intended to apply broadly, and each community is afforded the flexibility to identify the particular issues plaguing it. In New York City, rampant crime was just as significant a problem as it was in the subway, yet its causes and manifestations were by no means identical. The subway experiment is illustrative of broken windows' impact in a closed system. New York City's experience has proven to be more contentious, no doubt because of the innumerable factors that may be attributed to crime reduction in such a large municipality. Nevertheless, the transition of William Bratton to commissioner of the NYPD in January 1994 suggests that the theory was an integral component of New York's recovery strategy.

As a young child, I remember driving with my family through the city (we lived in the suburbs, approximately thirty miles outside of New York City) and being a bit nervous of highway mergers and intersections. At the time, the last vestiges of "squeegeeing" were beginning to be curtailed. Squeegeeing is the "unsolicited washing of car windows," often by groups of washers approaching cars stopped at red lights or in heavy traffic by the entrances of tunnels, bridges, or mergers. While it sounds innocent enough, watching a crowd of men prowl between cars was a disquieting experience for a young boy. It was also unsettling for adults, as it turns out. Some squeegee-men who were denied compensation would fold themselves onto the hoods of cars to obstruct their passage, others would spit on windows or spray grimy water on them to obfuscate a driver's vision, and occasionally vehicles might even be swarmed by a gang of squeegee-men. Just as Bratton and Kiley had been trapped in subway fare extortions, so too did then-NYPD commissioner Raymond Kelly have his car serviced and subsequently spit on as he and his wife drove through the city. Soon after, Kelling was contacted to help curb the pandemic.[42]

Squeegeeing was punishable by a "desk appearance ticket" (usually a fine), often called the "disappearance ticket" because those served rarely made a point of keeping their appointments. Missing a desk appearance resulted in the issuance of a warrant, but those warrants were almost always lost in the flurry of outstanding warrants on more serious crimes. Police found themselves helpless in the face of obstinate offenders who were outside the law. Eventually a street officer came to the commonsense realization that if patrolmen could access warrants themselves, they could arrest previous offenders on their beat and bypass the warrant-service unit entirely. In a matter of weeks, squeegeeing was almost eradicated as growing numbers of arrests demonstrated police commitment.[43] Through a focus on small-scale crimes, signaling a renewed emphasis on order maintenance, broken windows policing had made its debut on the streets of New York and proved decidedly effective. Concurrently, in January 1994 William Bratton was appointed police commissioner. As in the subway, Bratton's tenure saw an immediate spike in attention to quality-of-life crimes.[44] One and a half years into his stewardship, a city so recently overrun with crime had changed enough to declare "The End of Crime as We Know It" on the cover of *New York* magazine.[45]

In 1992 New York experienced 2,154 murders and a total of 626,182 index (or major) crimes. Five years later, murder rates were down 64.3 percent (to 770) and total index crime had been cut in half (355,893).[46] By way of contrast, in 1994, murder rates had declined 5 percent nationally, while they far exceeded that mark at 17 percent in New York.[47] As he did with the subway, Bratton ensured that enforcement focused on perceptions of disorder, this time at a local level. High-profile weekly meetings between top chiefs and precinct commanders ensured that members of middle management were made aware of their accountability to the broken windows approach; misdemeanor policing became the dominant means of conveying a sense of control and presence on every block in the city. And just as in the subway, it soon became obvious that low-level crimes were often perpetrated by violent offenders or by those with knowledge of such offenders (crime's multidimensionality). Kelling and Coles provide several examples, including how someone arrested for public urination, upon further interrogation, provided police information on a totally unrelated and more serious crime (knowledge of a stash of illegal weapons). In another example, someone selling illegal goods, upon greater scrutiny, provided police with information about an illegal weapons operation.[48]

Of course, not every minor incident led to such a find. Recall that in the subway, arrests of fare beaters only turned up a weapon or an outstanding warrant one time in ten. Yet it was enough to turn the tide. Broken windows policing allowed for the collection of "accurate and timely intelligence" by encouraging officers to interact with a subset of the criminal population they long ignored; local discretion allowed responders to follow up on the intelligence they gathered quickly, and proper oversight ensured that such discretion was employed appropriately.[49]

There is no doubt that myriad other factors were involved as well. I noted that crime rates across the country were shrinking, if not as dramatically as the rates in New York. In accordance with Gladwell's argument, Bratton, Kelling, and Sousa suggest that broken windows policing amplified a variety of forces to create a tipping point in the fight against the crime epidemic, adding compelling leverage to preexisting trends.[50] Yet, as we saw in the subway example, the broken windows theory's implementation can in fact create a tipping point in and of itself. We could also look at the example in New York City, not as a single case but as dozens of related but distinct studies. While New York is only one city, Kelling and Sousa note that each of its seventy-six precincts is in truth a city unto itself, with distinct cultures, socioeconomics, and populations exceeding 100,000. Each of these precincts received the same policing style under Bratton, with the ability to tailor broken windows to their specific needs, and seventy-four out of seventy-six (more than 97 percent) did not differ significantly with respect to crime rate reduction.[51]

The full story of New York has been outlined in detail in Kelling and Coles' *Fixing Broken Windows*, Gladwell's *Tipping Point*, Bratton's *Turnaround*, and other works. The major lessons for this research can already be discerned without further detail. Both New York City and the New York City subway system illustrate the effectiveness of order maintenance as a means of violence reduction and prevention. Moreover, the theory's quick implementation by the TPD and NYPD demonstrates that broken windows can help retool large organizations to nimbly and judiciously enforce against pervasive disorder, employ timely intelligence, and interrupt the cycle of disorder and violence that broken windows would contend contributes to instability in hotspots across the globe.

In *Learning to Eat Soup with a Knife*, John Nagl argues that for an institution to learn and evolve, the rank and file must have an opportunity to transmit their experiences upward. Subordinates, he notes, must be encouraged to question policy, offer suggestions from the field, interact with higher ranks, and inform informal procedures rather than have them bureaucratically imposed from the top down. David Kilcullen, in pondering potential lessons that militaries might glean from police and emergency services, specifically identifies how successfully the latter organizations push command decisions down the chain.[52] It is no surprise, therefore, that broken windows policing has been successfully applied with such a management style. Bratton is renowned for his ability to restore order not just to cities (he was police commissioner in Los Angeles and Boston as well) but also to police departments.[53] Upon taking the reins of the NYPD, he immediately devolved power from a handful of chiefs down to the seventy-six precinct commanders because true order maintenance requires an intimate familiarity with local context and custom.[54] Kelling and Coles posit that law enforcement activities may only comprise about 20 percent of the daily workload of a police officer walking his or her beat. The other 80 percent of his or her time is spent on unofficial order maintenance—circumstances in which police use discretion to determine not only whether a crime has been committed but also whether the behavior at issue requires formal censure or extralegal responses. This process necessitates the exercise of discretion on the part of officers, which anticipates that the daily routine of police cannot rely on "rote application of specific rules, but on the application of general knowledge and skill."[55] There is a degree of flexibility in how police respond to factually identical situations, and chiefs have come to accept, borrowing military parlance, the "war being fought by junior commanders,"[56] handing them "the responsibility to make decisions on the spot."[57]

Marine Corps lieutenant general Victor "Brute" Krulak said, "You cannot win militarily. You have to win totally, or you are not winning at all."[58] Yet, while navies have long theorized about their roles in peace and war, the gray spaces in between are far less conspicuous in strategic literature. "The American way of war" has historically been "marked by a belief that the nation is at war or at peace; [a] binary nature of war" that resists the recognition that a soldier likewise may spend less than 20 percent of his or her career fighting "real" wars.[59] Further, it seems fundamentally hard to know when states are at war and when they are at peace, perhaps more so now than ever. Protracted and inconclusive conflicts in Afghanistan and Iraq, alongside instability across the Middle East, North Africa, the Caucasus, and Central America, demonstrate both the need and the notable lack of a strategic lens the Navy can use to make sense of a disordered post–Cold War order. In the twenty-first century, militaries have struggled to find a strategy that supports Krulak's heuristic, but broken windows has proven its relevance and utility in the gray spaces of order maintenance, winning back the confidence and safety of American neighborhoods as well as the accolades of those responsible for its implementation.

Broken Windows beyond Policing

While broken windows gained prominence as a criminological construct, this book is not the first time its theoretical underpinnings have expanded into other disciplines. Medicine, psychology, and sociology all engage with the powerful relationship that broken windows posits between context and behavior. And speculation about that interplay has existed for far longer than the theory itself. Earlier I noted the 1969 car experiment. In another study predating broken windows, academics at Johns Hopkins University established that, when it comes to neighborhood relationships and mental health, "space [i.e., context] in and of itself is an important factor."[60] Twenty years later, Kelling and Wilson capitalized on the validity of that hypothesis, and fields like public health have quickly made it their own. Papers on topics ranging from understanding self-esteem to the spread of venereal diseases incorporate aspects of criminology for a fresh look at persistent, chronic issues. In one study, partially funded by the Centers for Disease Control and Prevention, Louisiana-based researchers analyzed the gonorrhea rates of 26,600 people across fifty-five block groups (a census term) in the city of New Orleans.[61] Using digital mapping features to plot all reported cases of the infection (reporting is mandated by law) and visual data collected by the University of New Orleans on street conditions in each block group, the researchers were able to construct a broken windows index to quantify the degree of disorder on a given street and compare it with the spread of venereal disease.[62] The methods employed make it a telling example of how Kelling and Wilson's theory has been adopted by the public health community.

The researchers proceeded to map the other variables often associated with the spread of gonorrhea, such as the presence of liquor stores, the racial makeup of each block group, the marital status of those in the groups, and their level of poverty. Of all the variables tested (using a least-squares regression), only the broken windows index and the poverty index showed any significant relationship with the prevalence of gonorrhea across blocks, with the broken windows index expressing a higher degree of relation with disease rates than poverty. In fact, orderly neighborhoods (i.e., those with low broken windows scores) with high rates of poverty showed no discernable difference in infection rates when compared to neighborhoods that were orderly but had low rates of poverty.[63] Environment, not income, mattered the most. Measuring the orderliness of a neighborhood was a greater predictor of gonorrhea prevalence than any previous variable used, including longstanding indicators such as race and socioeconomics. The authors do not go so far as to determine the nature of any causality that this correlation implies, but the results open a world of experimentation for the broader applicability of Kelling and Wilson's idea. The study even suggests a trial intervention program to better test the environment's direct impact on high-risk sexual behavior.

The Louisiana gonorrhea case study (who would have thought you'd be reading that line in a maritime security book?) focused primarily on a literal interpretation of broken windows, referring to the physical environment in which people live. It was the tangible setting of New Orleans and its relationship with high-risk sexual behavior that was the focus of examination. And for good reason—such a literal interpretation is easily observed and often produces useful propositions for practitioners (e.g., clean up the graffiti on a given street). Yet the concept of an environment can be interpreted far more broadly, to include a state of mind.

The seminary experiment from earlier in this chapter offers a good example of an expanded interpretation of context. It was the temporal context (being late or early to deliver a sermon) that impacted the actions of those seminarians, not something they could see or touch or hear. Studies that help forge a bridge between the more literal broken windows interpretation (physical disorder) and Gladwell's (and, indeed, Kelling's) broader application of the idea of context include those like one survey of youth safety across Baltimore, Detroit, Oakland, Philadelphia, and Richmond.[64] The outcome of the paper is, at this point, unsurprising. The authors found that feelings of fear at school among students were directly related to those students' beliefs that their schools were disorderly. Once again, most common demographic proxies appeared to be "less strongly associated with feelings of safety" when compared to feelings of disorderliness.[65] However, because the interviews with students were conducted by phone, the authors could not evaluate firsthand the physical disordered state of each respondent's school. Instead, the authors used the reaction to the statement, "Kids can get away with almost anything at my school" to code for disorder.[66] Because of this, the authors claim that the study is limited due to an inability to directly assess school conditions. This is, however, also an opportunity to better understand disorder.

Disorder, in the context of broken windows, is ultimately about how we feel about our environment and, consequently, how those feelings influence our actions.[67] This human component, whereby we internalize and interpret the signals in our environments, means that context is more than just the street and the garbage around us; it includes the less tangible signals, like behavior and time and sentiment. This human component also means that the way we react to environmental signals does not necessarily reflect objective reality. It is enough that we feel disorder that drives our behavior, whether or not one might dispassionately characterize an environment as disorderly. This is an emotional, constructed approach to the concept of disorder. For that reason, in the case of the students, how they felt about the state of their school was perhaps an even better measure of disorder than having researchers actually assess the schools in person. Contextualizing our surroundings, as these experiments demonstrate, is as much about how someone emotionally perceives their environment as it is about the physical appearance of that space. It was this wider notion of context that allowed for a nuanced application of broken windows to public health, to New York City's streets, and (in the coming chapters) to the maritime domain.

Room for Debate

While the broken windows theory has gained wide prominence, it is not without its shortcomings and detractors, a survey of which is important for an honest discussion of what the theory can (and cannot) tell us in the chapters to come. Fair warning, however, that this section gets a little in the weeds on the theory, and readers could safely skip to the following section if the details are not particularly pertinent to them.

Broadly, we can speak to two related types of mostly quantitative arguments directed in criticism of the theory, what I have categorized as "chicken and egg" and "too much noise" disputes. In the first, the chicken and egg, critics argue that it is unclear in what direction variables are acting on one another or if they are collinear. In the second, the too much noise argument, critics contend that the abundance of available variables (the noise) makes it difficult to identify which variables actually say something significant (cause versus correlation).

Robert Sampson and Stephen Raudenbush, for example, leverage the first argument to critique broken windows. They posit that disorder and violent index crimes (with the exception of robbery) are not linked in a causal chain or cycle but are instead both symptoms of some shared underlying cause.[68] In other words, the variables are collinear. Controlling for collective efficacy (a neighborhood's willingness to intervene in its defense) and structural issues like building codes, Sampson and Raudenbush suggest that disorder might be more accurately seen as part of a larger ecology of crime, not a causative factor explicitly resulting in criminal action.[69] As a consequence, they conclude, while patrolling disorder may have incidental effects on crime, a broken windows approach is inefficient and unfocused, targeting a symptom of the same underlying cause as crime.[70]

Broken windows proponents respond to the chicken-and-egg debate (is disorder causing crime, is crime causing disorder, or are both caused by something else?) by arguing that the theory is not invalidated when other indicators are shown to have some relation to crime. The theory's foundational premise, they contend, is that crime and environment produce a feedback loop, that crime builds on the presence of disorder. In the example at the top of the chapter, one broken window did not decidedly cause a murder; rather, the broken window precipitated a series of events that, over time, produced the environment in which a homicide might take place. There are opportunities all along that chain of events in which intervening (at increasingly higher cost) could have prevented such an outcome, beginning with fixing the broken window. Sung Joon Jang and Byron Johnson say as much in their critique of Sampson and Raudenbush.

Jang and Johnson argue that Sampson and Raudenbush misinterpret the process by which disorder leads to crime when the latter pair observe, "A fundamental thesis of 'broken windows' is that observed disorder directly causes predatory or 'serious' crime."[71] Jang and Johnson take issue with the word "directly" in this summary of the theory. They suggest, alternatively, that "disorder indirectly causes crime via weakened informal social control."[72] Casting disorder's role as indirect, serving to dampen a community's willingness or ability to police itself, Jang and Johnson reaffirm the "community" in community policing. More to the point, if we take Jang and Johnson's framing for how disorder fits into the cause-and-effect cycle leading to violent crime (i.e., that disorder acts indirectly, through the community), Sampson and Raudenbush's conclusion that controlling for collective efficacy reduces the impact of disorder on crime would actually be proof positive of broken windows.[73] In that instance, factoring out a neighborhood's willingness to police itself is actually controlling for the very effect to which broken windows speaks. Jang and Johnson are not unique in casting disorder as an indirect mechanism leading to more violent crime. Sousa and Kelling both argue, when they stated that disorder, if "left unchecked by community and neighborhood controls," would lead to serious crime, they were alluding to this indirect pathway.[74]

If that argument seems unsatisfactory, there have been efforts, such as those by Wesley Skogan, to more categorically answer the chicken-and-egg debate. Skogan posits that there is evidence of a statistically measurable causal relationship between community disorder and crime even when controlling for other factors.[75] In other words, one (disorder) can measurably be said to come before and cause the other (crime). Yet Skogan's findings are not universally persuasive, even among broken windows proponents. Some broken windows literature even leans in the opposite direction entirely, embracing a hypothesized process in which disorder and crime (or high-risk behavior, in the public health literature) create a feedback loop in which they mutually influence one another.[76] Ultimately, the debate over causation rages on. As a consequence, the argument for a clear, quantitatively proven causal link between disorder and crime remains one of the weak points in the literature.

Causation versus correlation is a related and particularly recurring disagreement in interpretations of broken windows' statistical models, and it seems that the sheer volume of available data only perpetuates this debate. Crime is an expansive topic and large cities create an impossibly long list of noise in which models can invariably be shaped to fit.[77] Bernard Harcourt and Jens Ludwig, for example, employ such an argument to discredit one type of broken windows research model by creating their own theory, the broken Yankees hypothesis.

The broken Yankees hypothesis suggests that when the Yankees baseball team does well, New Yorkers, united by a sense of common purpose and encouraged to bond with fellow bar and restaurant patrons, commit fewer violent crimes.[78] Conversely, when the team is doing poorly, residents of New York do not share a common goal and therefore crime escalates. Remarkably, there is some statistical evidence to suggest Harcourt and Ludwig are on to something, enough to make you think twice about the theory. Most likely, the broken Yankees hypothesis is an aberration; there is no proven connection between the two variables of baseball success and crime rates. In a city full of possibilities, Harcourt and Ludwig could choose from hundreds of accidental correlates with crime to make claims that seem valid. Pointedly, while those authors were using such an example to criticize evidence in support of broken windows, the larger conclusion may be that some problems are by nature too complex to model sufficiently.

Harcourt and Ludwig's use of baseball in their critique is instructive because sports are in many ways similar to this conversation. The obsessive collection of statistics on nearly every measurable baseball metric provides, like with crime, a trove of data to analyze. Yet, just as with crime, the abundance of information can make discerning coincidence from cause a difficult process. Nate Silver offers a similar example in his book *The Signal and the Noise*. From the first Super Bowl in 1967 through the thirty-first Super Bowl in 1997, he notes that the winning football team's conference appeared to be a leading indicator for stock market performance. If the winning team was from the original National Football League, the market would gain an average of fourteen points on the year. If the winning team was from the original American Football League, the market would lose an average of ten points on the year. Twenty-eight of the first thirty-one Super Bowls accurately predicted the following year's trend. If you took this marker seriously, you would calculate there was only a 1 in 4,700,000 chance that the relationship was merely coincidental. If you took this marker seriously, you would have lost your shirt starting in 1998. In fact, since then the results have almost been entirely reversed, with American Football League teams predicting a 10 percent growth and a National Football League team clinching victory just before the 2008 housing bubble burst.[79]

Perhaps the easiest criticism to make of broken windows is that it is counterintuitive. In a letter-to-the-editor forum in the journal *Environmental Health Perspectives*, one public health practitioner responds to the claim that gonorrhea rates could be connected to environmental disorder by saying, "My instincts tell me this would not be the case. The 'Broken Windows' are a consequence of the behaviors of that particular community and they are not the cause of the behaviors."[80] This is essentially a colloquial version of Sampson and Raudenbush's argument that disorder is a fellow symptom, not a cause. I have already addressed the broken windows counterpoint to that criticism from a quantitative standpoint, but there is a more fundamental point to address as well. The author speaks to instinct, which makes it difficult to judge that disorder could play (directly or indirectly) on our behavior. Yet, as the theory would contend at least, we are subservient to the signals bombarding us. Psychologists like Walter Mischel suggest that this disconnect between what we experience and what we think we experience happens for a reason: the human brain has a "reducing valve" that allows us to filter out the complexity in life.[81] Broken windows is not inherently intuitive, at least not in the magnitude it purports that we are influenced by the world around us. Without closing the reducing valve, the true relationship between disorder and violence is easy to miss. In New York and cities around the United States, it was little noticed for decades.

More powerful criticisms are often those advanced against the models or case studies used to support broken windows, not those directed at the theory at large. The broken Yankees hypothesis was one such example. Another appeared in a paper investigating the cause of the dramatic downturn in homicide rates experienced in New York City.[82] That study suggests that dividing homicides by race would allow for a more accurate picture of causation. Factors such as firearm availability, changes in drug use, and increased incarceration rates were subsequently compared to racially sorted homicide statistics. The results show that decreases in African American homicides were strongly associated with decreases in crack cocaine consumption, Hispanic homicides were tied more to gun availability, and white homicides saw no strong predictive indicators among the variables tested. The authors conclude from this data that the cocaine hypothesis on crime reduction is supported, whereas the broken windows theory is not.[83]

This is not altogether an inaccurate conclusion. Gladwell, in particular, agrees that the dip in cocaine use over that period shows a relationship with decreased violent crime.[84] That dip, however, was a nationwide trend in the 1990s, as were positive developments in the economy and changes in demography typically associated with lower crime statistics. The extraordinary collapse in crime rates in New York should therefore parallel that of other major cities if those indicators were the dominant causes of crime reduction. Yet they do not, nor do they explain the speed at which criminal behavior diminished in New York as compared to other cities.[85] This is all the more noteworthy because New York City's poorest had been hit hard with welfare cuts in the 1990s, and national aging trends (associated with lower crime) were not mirrored in a city that was actually getting younger. With respect to drug use specifically, cocaine had been in steady decline well before crime rates in New York showed signs of wavering.[86] The study may obscure this confusion of long-term trend with short-term effect as a result of its analysis of the decade as a whole, whereas New York homicide rates dropped 64.3 percent in only half that time.[87]

More important, the depiction that African American, Hispanic, and white homicide rates are associated with different root causes does not preclude a broken windows approach to crime reduction. If anything, the variety of causes that precipitate crime is an endorsement of the complications of tackling violent crime only at the root cause and not the broader, contextual level. One could easily have said in the 1990s that police simply needed to target cocaine consumption to decrease African American homicides. And surely they should have. But how many factors contribute to cocaine usage and dealing? Does that vary by race as well? Does it vary by socioeconomics? And how many factors contribute to each of those factors? The inherent multidimensionality of crime makes it difficult to make short-term progress through an exclusive focus on root causes, even as they are sure to remain an important (and growing) part of the conversation. What the broken windows theory shows is that there is another powerful medium through which crime can also be addressed. "The strength and direction" of all other variables is "always dependent on the local context," and that is something police and their policies can noticeably impact.[88]

Unfortunately, this is not a debate we can resolve here with any finality, and the finer points of the debate are better left to the criminologists. What we can take away, both from the public health and criminology literature, is that a broken windows approach places context and multidimensionality at the center of human behavior, prompting unique and actionable insights for policymakers, insights with potential utility when combating insecurity in the littorals as well.

Putting It All Together

To understand how the broken windows theory relates to several other relevant spheres of research, it is useful to disaggregate the theory into three components. First, broken windows is a theory of order, norms, and social behavior. Second, the theory is one of crime, particularly how crime proliferates and is experienced at the community level. Third, it is a theory of policing and how to mitigate crime. Triangulating broken windows within these three streams of research provides a final means for placing this book within a broader framing, from psychology to critical studies on security and to wider debates in criminology.

The first component of this triangle—the social psychological dimension of broken windows—is seen throughout this chapter. Philip Zimbardo's 1969 car experiment provided clear, early evidence of the interaction between neighborhoods and context. Zimbardo's experiment highlighted an emerging notion of how communities interact with their environment in the construction of social norms and, significantly, how easily disorderly signals can fracture those norms. This can also be seen in Wilner and colleagues' 1960 study on the importance of the physical environment in understanding mental health.[89] Of interest as well is Zimbardo's infamous Stanford Prison experiment, conducted in 1971, in which randomly divided groups of "inmates" and "prisoners" adopted their hypothetical personas to frightening ends. The experiment served to reinforce Zimbardo's earlier hypothesis that actions (both at the community and individual levels) are guided not simply by internal rational considerations but by external stimuli and perceptions.

Zimbardo's prolific experiments have sparked whole fields of study and represent a good starting point for those interested in investigating the psychological underpinnings of broken windows. The theory relies on a wide body of work that lays the foundation for understanding the role, construction, and destruction of norms with respect to order and social behavior. Beyond Zimbardo, I also noted an array of psychological studies that reinforce and inform broken windows—take those on seminary students or on grade school cheating, for example. Each of those cases demonstrates, from a psychological angle, the direct relationship between perceptions of context (both physical and behavioral) and actions. Finally, Mischel's concept of the mind's reducing valve—wherein we filter out complexity to aid in decision-making, occasionally leading to oversimplifications—helps us understand how we might not consciously recognize all the subtle ways in which our environment shapes our behavior.

Second, the broken windows theory looks to understand what relationship exists between this understanding of norms and the world of crime. What results, as I have shown here, is a portrait of how small-scale disorder erodes communal self-efficacy and leads to more violent crime over time. In the example at the start of this chapter, I sketched how a single "broken window" signals that a neighborhood is uncared for. In the example, this precipitated a flight of families and the degradation of the community's capacity to self-police. As the neighborhood became ever-more disorderly, small-scale crimes escalated, perpetuating the cycle of criminality and the perception that the community is abandoned. Scaling up Zimbardo's groundbreaking work to the urban scale, norms can be seen to play a fundamental role in how crime rates rise and fall.

One particularly relevant spinoff of broken windows is the literature on everyday security. Everyday security is a critical study of security that aims to contextualize power, which is exercised by the state, as average people experience it. Like my study here, everyday security places a premium on how communities perceive the state's policing efforts and how that in turn shapes their relationship with their own neighborhoods (and police). As Adam Crawford and Steven Hutchinson describe the theory, everyday security offers two objectives: "drawing attention to both (1) the ways in which security projects are experienced, and (2) how individuals and groups deploy certain practices to govern what they understand and interpret as their own security."[90] Crawford and Hutchinson even focus on some of the dimensions of disorder seen throughout this chapter, notably the "temporal, spatial, and emotional features" that inform security.[91]

Understanding how security tactics are experienced at the community level is part of a rich tradition in the literature on critical studies of security. The University of Texas' Ben Brown's focus on policing methods, for example, dovetails well with Crawford and Hutchinson's approach. Brown writes of the ineffectiveness of increasingly invasive police operations and, of most relevance here, the counterproductive resentment that combative tactics can produce within communities.[92] This emphasis on the lived experiences of policed communities also intersects with research on cultures of insecurity. Marie Breen-Smyth, for example, addresses not how we construct policing, but how security forces construct the community. She writes to the risk of marginalizing populations by categorizing them broadly as suspect communities.[93] Breen-Smyth details, through a powerful autoethnography, the self-defeating repercussions of the impulse for police to construct and target "others." By placing regular people in the foreground, everyday security—and critical studies more broadly—exposes the vulnerabilities that communities face when policed by the state.

Third, armed with an understanding of how norms shape communal disorder and how a breakdown in those norms perpetuates crime, broken windows ultimately arrives at an understanding of what that means for police. It means being attentive to community concerns and focusing on interrupting the cycle of escalation at the "softer" end of the criminal spectrum (e.g., squeegeeing or fare beating). I referred frequently to these acts earlier as signals of disorder. Earlier I also spoke of the importance of emotion in constructing our environment, that how we *feel* about our environment is more important than any objective reality. Martin Innes dives deeper into these signals in work that has been characterized as a "natural successor to Wilson and Kelling's 'Broken Windows Theory.'"[94] Innes capitalizes on this relationship between disorder and fear to better understand how some crimes generate greater fear in communities than others. These aptly named "signal crimes" help explain why neighborhoods may *feel* as if crime is getting worse even if overall rates of criminality decrease.[95] Throughout this book I return time and again to the importance of how communities internalize their environment and what this means for littoral security.

Geoffrey Till has written of maritime security that the newly popular term may simply be a rebranding of the more conventionally understood notion of good order at sea.[96] However one chooses to label the concept, good order at sea lacks a library of theories upon which scholars and policymakers can draw. That Till dedicates so much space to order maintenance in his seminal work on sea power is confirmation of maritime security's growing importance on the naval strategic stage. That Till's chapter on the topic is particularly theory-light, however, is confirmation of the work still to be done on developing frameworks of good order. Till's chapter, robust as it is, ultimately relies on familiar mechanisms for discussing maritime security—specifically, through an evaluation of individual threats that ultimately relies back on the conventional counterthreat narrative. It is my hope, as we proceed to the case studies, that the broken windows theory and its component parts—its aggregated theories on norms, crime, and policing—provide a useful framework to operationalize what we mean by good order at sea.

Approach

The remainder of this book, as discussed in the introduction, is divided into two parts, one large case study (Cocaine and Context in the Caribbean) and two smaller comparative cases (Integrating Piracy). This approach is designed to meet the two demands for these parts—theory building and theory testing, respectively. Insofar as the broken windows theory has never been employed in a maritime context, this book plays an inevitable role of building a theory. For this type of task, robust single case studies are common tools of analysis. However, insofar as criminologists have already extensively debated the theory and its internal mechanisms, this book doesn't fit exclusively in the world of theory building. Instead, it also plays a theory-testing role, taking existing concepts and stressing them in new ways. For that type of analysis, cross-case comparisons are typical tools. So here I fuse both approaches in one book. Part 2, on the Caribbean, represents a single, in-depth case study to build our understanding of how the broken windows theory might be deployed in the maritime domain. You can think of this as providing a proof of concept. Then, using the language and ideas constructed in that case, the following part offers a cross-case comparison to test those ideas. In that comparison, we'll see whether the theory is generalizable to new issues and new places. In choosing cases, I aimed to select littoral areas that the strategic community already recognizes as distinct regions with distinct literatures. These would include the Caribbean, the Mediterranean, the Gulf of Aden, the Straits of Malacca and Singapore, the Gulf of Guinea, and the Gulf of Mexico, for example. Narrowing the selection to the three cases I ultimately chose was largely a function of finding good data, looking for regions with persistent levels of maritime instability, and avoiding regions with serious failed state problems.

One of the most important considerations when deciding what types of cases I would use in this book was the need to facilitate the scaling up of the broken windows approach from its urban roots to the maritime space. How do you bridge that divide? One way is to look at both cities and maritime spaces as systems. As one paper on urban systems notes: "Cities, urban areas, and urban sprawl are multidimensional systems comprising individuals, communities, society, and economy, in a common geographical location where different types and degrees of interaction occur."[97] Kilcullen similarly depicts coastal cities in systemic terms.[98] Discussing cities as systems means we explicitly understand that urban areas exist as relatively distinct geographical and sociological units. As I discussed in relation to the New York City subway case, when broken windows is placed in this context, we can clearly discern that the theory's precepts are not tied uniquely to an urban environment but rather to an ecology (system) of human interaction. A logical escalation, in my mind, is to go from urban to regional. Regions, like cities, are marked by distinct literatures that identify them as self-evident systems. Like cities, regions are of course not perfectly closed. The inherent artificiality of isolating a city from its suburban and hinterland roots is equally at issue in isolating one region from that of an adjacent one. That both cities and regions are recognized units of sociological study, however, suggests that regions can be fairly treated as systems of their own and thus may make good parallels for the urban environment at a larger scale.

A second obstacle to scalability comes in understanding the demands on data. My goal in the coming chapters is to see if some of the key elements identified above (context and multidimensionality) are present in the cases we explore. Doing so would suggest whether the broken windows theory is in fact a useful tool for thinking strategically about maritime security. This analysis requires a lot of descriptive data, which resource constraints meant I was not going to get on extensive field investigations. And, simply put, regions are often the most consistent levels of analysis for which transnational data (on issues like narcotics or human trafficking) is available. This demand for descriptive data is even more difficult to satisfy in the maritime context, where greater geographic distances and imperfect reporting make data collection at smaller levels exceedingly difficult. Thus, while case selection at the regional level is informed by a methodological need to bridge the broken windows theory from the urban to the maritime, it is also born of a practical consideration when assessing how best to find data for that enterprise.

❖ ❖ ❖

In this chapter, we teased out two critical lessons learned from our exploration of the broken windows theory of crime. First, crime is informed by environment, and shaping environment may provide a means of moderating crime. Second, broken windows embraces the multidimensional nature of disorder—crimes are not isolated phenomena, and policing small-scale disorder pays dividends along the escalatory crime scale. Such a review also highlights several unresolved issues in broken windows research, particularly the role of establishing a statistical causal relationship between disorder and index crimes. While it is clear that broken windows is not a perfect theory, its precepts of the importance of context and the role of multidimensionality provided an actionable lens by which police could understand and combat crime in major cities. I contend, as we will explore in the following sections, that the same is true for navies in the littorals. So, let's put that to the test.

PART TWO
Cocaine and Context in the Caribbean

The *Cyclone*-class coastal patrol ship USS *Zephyr* (PC 8) transits the Caribbean Sea. *Zephyr* is under way in support of Operation Martillo, a joint operation with the U.S. Coast Guard and partner nations within the U.S. Fourth Fleet area of operations. *U.S. Navy photo by Mass Communication Specialist Seaman Kenneth Rodriguez Santiago/Released*

CHAPTER THREE
The Business of Drugs

THE REGION

Counternarcotics, from a narrow perspective, is the effort to eradicate the sources of drug production and dismantle the networks that profit from their distribution. At the broadest level, however, nearly every act of law enforcement and national security in the Caribbean is a form of counternarcotics policing. Money laundering, terrorist financing, drug abuse, homicide, forced prostitution, irregular migration, and perceptions of failed governance are all interconnected elements of the smuggling and sale of drugs, as we will see in part 2. From the Caribbean's permissive attitude toward lax banking regulations, to government corruption, to local crime, to the illicit international trafficking in people and weapons, the movement of narcotics through Caribbean communities is made possible by and contributes to a much broader regional maritime insecurity. Thus, even while efforts at countering the trade in drugs have been redoubled since the temporary withdrawal of resources after 9/11, a truly effective maritime counternarcotics strategy is one that places drugs within this wider, multidimensional framework. Doing so, however, requires an understanding of how these myriad crimes and disorders interact. In the ensuing chapters, I highlight this interaction. By directly observing the multidimensional nature of Caribbean instability and placing crime in a broader context, we take the largest single step in demonstrating the applicability of broken windows as a theoretical foundation for a strategy of maritime security.

Throughout part 2, it is important to remain cognizant of the primary lessons established in chapter 2. There we discerned two themes in our discussion of the broken windows theory—that crime is context-dependent and multidimensional. First, at the social psychological core of the theory lies the expectation that rational actors only behave rationally within a given set of restricting parameters (their context). Human behavior is inexorably tied to how we contextualize and internalize our environment. Malcolm Gladwell illustrates just this point with the power of context, calling our attention to experiments conducted with populations from seminarians to grade school students. Likewise, signals of government abandonment and community neglect act on the social psyche and create conditions in which high-risk behavior (unsafe sex, drug abuse, crime) is more likely. Because of the spotlight this shines on perceptions of insecurity, broken windows policing places a premium on a community's sense of safety. By focusing on a population's attitude, community policing can impact violent crime rates in part because it increases a community's willingness to act in its own defense (self-efficacy).

The second lesson, multidimensionality, comes when we evaluate the real-world applications of this theory. It became evident, in both the New York City subway and the streets themselves, that addressing the seemingly small gripes of the city's residents (fare beating or window washing) had compounding success on crime prevention. This was the case in part because of the signal that such initiatives sent to the public, that something new and responsive was being done to combat crime. Yet practical experience taught that there was an added element to the success of cracking down on lower-order threats. It soon became apparent that those who committed petty crimes were often the same individuals, or knew of those individuals, who committed more serious ones. Policing a city's broken windows proved effective not only for its hypothesized social psychological component but also because of the operational impact of placing line officers in direct contact with those most at risk.

Together these themes support a perspective on security that is multidimensional and context dependent. Moreover, the themes suggest a vision of security that places the most pervasive and endemic issues (however "small") at the top of the priority scale precisely because they provide the daily context for the vast majority of people. One of the broken windows theory's signature emphases is on problem solving at the community level. By taking seriously the "quality-of-life" crimes that plagued New York, the New York Police Department helped stabilize a city gripped in a perceived epidemic of violent crime. The same can be true on the world stage. Homicides in the Dominican Republic, Honduras, and Puerto Rico may seem distant to the average American. Yet, as we will see, rising crime rates in these countries are related to the widespread trade in small arms and light weapons across the Caribbean Basin, which in turn is connected with narcotics smuggling, human trafficking, and even terrorism. No element of Caribbean security can be understood outside of this multidimensional context. When police treated violent crime as a symptom of disorder, combating context over crimes, they quickly learned that securing community attitudes often meant impacting crime on a wider scale. The same principle applies on a global level. An American maritime security strategy guided by the broken windows theory emphasizes the endemic security issues that shape regional context, just as squeegeeing or fare beating shaped perspectives in New York. The broken windows theory is fundamentally one of context, and the strategic applications it engenders therefore necessitate a shift in the very way we frame security.

To help assess the dynamic nature of Caribbean insecurity, part 2 explores transnational crime in its myriad facets by analyzing how illicit networks and flows interact and what this means for a broken windows approach. First, however, we need to get a lay of the land, and this chapter is dedicated to that purpose. In the Caribbean, the most obvious issue is that of narcotics trafficking, and so it is there that we begin.

NARCOTICS

The Caribbean Basin is threatened by powerful currents of environment, economics, and crime. Such a multidimensional dynamic of stability, security, and sovereignty is not new to those who live and govern in the region. As former prime minister Lloyd Erskine Sandiford of Barbados notes, risks to Caribbean states are multifaceted. Islands are at physical risk of extreme weather; they are at economic risk, held hostage to larger forces beyond their control (including a reliance on tourism dependent on the economic winds of Europe and North America); and they are at political risk to terrorism and criminal elements.[1] Hybrid threats dominate the Caribbean, not easily characterized as internal or external in origin. For that reason, research on the security of small states has evolved over the past few decades away from purely state-level power relations. A focus on internal threats to stability—both as stressors to external threats and as issues of substance in their own right—has gained ground on the more conventional emphasis on external issues.[2] This sentiment has become such a staple in discussions on Caribbean security that at least four authors in one anthology alone, Ivelaw Griffith's *Caribbean Security in the Age of Terror*, pointedly emphasize the multidimensionality of small state defense.[3] The range of social, political, criminal, and economic obstacles facing Caribbean states compelled them to view security from this broader perspective.[4] Another Barbadian prime minister, Owen Arthur, summarizes the regional attitude as follows: "It would be a fundamental error on our part to limit security concerns to any one area while the scourge of HIV/AIDS, illegal arms and drug trafficking, transnational crime, ecological disasters, and poverty continue to stare us in the face."[5]

While such a perspective is common among Caribbean law enforcement practitioners, politicians, and local academics, it has made only a small impact on security strategies across the Basin. In a region fractured by so many small states, international cooperation often comes at the behest (and expense) of an external patron.[6] And while the region's chief patron, the United States, has invested heavily in multinational institutions and capacity building, the United States has struggled to embrace a regional security outlook matching that of local actors. This was exacerbated by the United States' seismic shift in priorities following the events of September 11, 2001. In the months and years after the attacks, the United States executed a massive transition of resource allocation, placing the greatest emphasis on counterterrorism operations. With limited deployable resources, much of that shift was manned and supplied by American assets from the Caribbean. Three-quarters of Coast Guard cutters and related aircraft, heavily drawn from the Caribbean, were tasked to counterterrorism responsibilities such as critical infrastructure protection for ports and tankers. Half of the Coast Guard's federal agents responsible for combating drug trafficking were reassigned as air marshals. The Federal Bureau of Investigation moved approximately four hundred agents from counternarcotics to counterterrorism missions.[7]

From the perspective of the era, this divestment of Caribbean resources devoted largely to drug trafficking was an easy choice. The Caribbean core, apart from the rest of Latin America (and excluding rim states like Colombia and Mexico), was seen to face virtually no direct threat from terrorism. While South and Central America ranked high on tables of international terror attacks in the years leading up to 2001, the Caribbean area in particular remained largely off the radar.[8] In an American context defined by the threat of internationalized terror, a region at no perceived risk or relation to terrorism could afford to cede resources. This was supported even further given the State Department's *Patterns of Global Terrorism 2002*, in which none of the organizations other than the obvious FARC and Shining Path (the Communist Party of Peru) appeared to have any overt connection to drug trafficking in the Caribbean.[9] In spite of a long-theorized relationship between terrorists and narcotics traffickers, the lack of apparent evidence to that point doomed the region to irrelevance in an American conception centered on terrorism.

Yet such a narrow perspective on security is exactly what Caribbean experts have long argued against. While responses to insecurity (crime, terrorism, piracy, etc.) may be atomized and bureaucratized, the nature of political violence is infrequently so well defined. The choice between counternarcotics and counterterrorism is artificial, as the UN International Narcotics Control Board argued when it urged the United States and others not to dip into counternarcotics to pay for counterterror but to find ways to gain efficiencies and combine efforts against both.[10] In that spirit, this section ultimately seeks to place a variety of security issues endemic to the Caribbean into a related framework. In so doing, I demonstrate how the broken windows theory helps us conceptualize America's interest in Caribbean defense in a manner that ensures human, national, and regional security and avoids false choices on enforcement options.

Background

The Caribbean's geographic location has long been among the region's most lucrative assets. For centuries, goods and people have moved between North and South America, between Africa and the Americas, and between Europe and the Americas, all passing through the Caribbean Sea. Since the completion of the Panama Canal, that traffic expanded to include commerce and migration from Asia and the Pacific using the Caribbean as a transshipment point for destinations farther afield. The Caribbean Basin is both an important maritime junction and home to one of the world's largest chokepoints, the Panama Canal. The Florida Strait, the Mona Passage, the Windward Passage, and the Yucatán Channel are all busy sea lanes crossing the Caribbean to join Atlantic, Pacific, and Gulf of Mexico shipping.[11] Today the Caribbean remains at the "vortex of the Americas," an intersection between North, South, and points farther afield. And while the geopolitical significance of this intersection was at its most dramatic during the Cold War, the subtler role of the Caribbean in contemporary politics and conflict belies its import.[12] A major part of that story is the value that nonstate actors, drug cartels in particular, have placed on the Caribbean for more than fifty years—"not in terms of geopolitics, but geonarcotics."[13] The very concept of geonarcotics, the internationalized nonstate distribution of illicit drugs and precursor chemicals, was made possible largely by the Caribbean's distinct geographical identity as a cluster of small and comparatively weak, resource-poor nations at the world's crossroads.

The drug trade is, at its roots, a business venture. It grew to meet the scale of demand emanating from the West, most heavily the United States. As postwar disposable income met baby boomer adolescence, the United States became an increasingly lucrative market for narcotics. As explained by Richard Millett, a Latin American specialist at the University of Missouri–St. Louis, through the early 1970s, a growing percentage of American-smoked marijuana originated in Colombia's Guajira Peninsula, an isolated jetty thrust into the southern Caribbean Sea. As the decade progressed, so too did the American palate. By the end of the 1970s the American appetite for drugs was moving decidedly toward cocaine, and Colombia followed suit.[14] By 1979 Colombia had already begun to structure its control over the flow of cocaine. At the time most coca cultivation remained in Peru and Bolivia, but processing and trafficking were increasingly centered in Colombia, with its access to both Caribbean and Pacific littorals. As cultivation was consolidated in Colombia, the country concurrently gave rise to the most infamous (and romanticized) personalities in drug lore. Reminiscent of the public fascination with Chicago's mobsters or New York's Mafioso, the 1980s witnessed the prominence of men like Pablo Escobar Gaviria, who in many ways continue to shape conceptions of what a drug don looks like. At the helm of the Medellín Cartel, Escobar rose to become one of the wealthiest criminals in history.[15] Yet Escobar's larger than life, self-styled Robin Hood persona belied a series of global shifts that would inevitably dislodge Colombia's organized crime networks from the pinnacle of the drug trade. Escobar was killed in 1993. His death provided a brief moment of opportunity for the rival Cali Cartel, though they themselves were brought down soon thereafter. Colombia would continue to remain the major global source for cocaine, but the trade's center of gravity was migrating elsewhere. At the heart of this shift was geography—or, rather, geonarcotics.

The closer cocaine (or any drug) comes to an American consumer, the greater its value becomes. For this reason, many drug smuggling operations in the Western Hemisphere set their sights on the United States. And they have been unreservedly successful, though the tactics they employ have varied considerably over the decades. By the end of the 1970s, James Zackrison, a former Latin American specialist for the Navy, recounts, "Entire fleets of speedboats lining up to offload drugs from mother ships were the norm."[16] Fishing boats and coastal freighters served as inconspicuous hubs from which so-called go-fast boats could load or offload cargo stashed in coves and safe houses. Yet the free-for-all elicited a considerable American response. By 1982, according to Zackrison, the South Florida Task Group had forced traffickers to transition from mother ships to aircraft, dropping payloads for elusive speedboats to collect and conceal. As cocaine distribution continued to grow in complexity and competency in the 1980s, American law enforcement presence similarly continued to build. In an effort to evade the tightening grip, traffickers transitioned away from shuttling narcotics between the Bahamas and South Florida (the preferred avenue at the time) and increasingly turned to the nexus of Puerto Rico and the Lesser Antilles. Yet American law enforcement was determined to close the Caribbean corridor. Radar placements in the Mona Passage (between Puerto Rico and the Dominican Republic) and Anegada Passage (between the British Virgin Islands and the Lesser Antilles) made the use of aircraft increasingly risky. Another blow was dealt in the early 1990s, when the American antidrug effort went multinational. The combination of American enforcement and growing regional opposition to the traffickers brought a decline in the popularity of the Central and Eastern Caribbean Sea smuggling lines, at the time the "lowest cost supply route."[17] Yet demand for a range of narcotics persisted, and so did the trade. In search of new avenues, the most obvious alternative became crossing the U.S.–Mexico border.

The rise of Mexican cartels was a direct result of Colombian criminal syndicates' efforts to more effectively deliver their products to market. The transition in power became a matter of fact when Colombian networks began paying Mexican smugglers in cocaine instead of cash.[18] With this shift in compensation, Mexicans became investors and wholesalers instead of merely middlemen and cutouts. The American border is the "single most lucrative bottleneck in the drug supply chain, the point where the most value is added," and the Mexican criminal groups that operated on the border (mostly smuggling people at the time) "possessed the most enviable situational advantage of all: territorial control of the approaches to the U.S. border."[19] Soon Mexican cartels had displaced Colombian organizations as the premier movers in the region. As with the Caribbean Sea route, this shift prompted heightened enforcement efforts on the United States' southern border. The American government began to emphasize efforts against Mexican distribution schemes, notably "the Arellano Felix organization in Tijuana, the Carillo-Fuentes organization in Juarez, and Cardenas Guillen in Tamaulipas and Nuevo Leon."[20] The increased presence once again highlighted the flexibility of the drug trade. This cat-and-mouse game is commonly referred to as the balloon effect, when enforcement efforts in one region or domain simply open enforcement gaps in another. This effect was on full display in 1993, when President Bill Clinton's antidrug directive produced a considerable shift in trafficking routes through the transit zone but precipitated no accompanying shift in drug use or sales. Before the directive, as Zackrison notes, as much as 80 percent of cocaine shipments had entered the United States through Mexico. After, as much as 80 percent of American-bound cocaine was once again delivered via the Caribbean Sea.

Today the Caribbean remains a vital link in the distribution of narcotics. In fact, despite the decline in public fascination, this century's kingpins (like Chapo Guzman) are more prolific in their sale of narcotics than their infamous predecessors ever were. Part of the reason for the persistence of the Caribbean smuggling route is the exponential increase in sale price of cocaine as it approaches American consumers. The Sinaloa Cartel, the cartel Guzman headed and the largest in Mexico, provides an illustrative example. That organization could buy a kilo of cocaine at the source for about $2,000. As that kilo moves closer to the highest valued markets (in this case, the United States), its price soars from $10,000 in Mexico to $30,000 wholesale (and a whopping $100,000 retail) once in the United States.[21] While the figures vary over time and by transaction, the general principle holds true whether that kilo flows through the U.S.–Mexico border in a false panel on a truck or a hollowed-out fuel tank on a fishing trawler heading for Miami. These combined efforts support a multibillion-dollar industry as lucrative and logistically complex as any licit global enterprise. The proceeds, however, are notoriously difficult to assess.

The U.S. Department of Justice places the revenue from American sales of cocaine for Colombian and Mexican networks anywhere between the wide range of $18 and $39 billion annually. The RAND Corporation provides a significantly reduced assessment, hypothesizing that Mexican cartels see revenue closer to $6.6 billion.[22] Even with the more conservative figures preferred by the *New York Times*, the Sinaloa Cartel, which controls between 40 and 60 percent of the cocaine market, would net approximately $3 billion annually.[23] Gen. John Kelly, then commander of U.S. Southern Command (later secretary of homeland security and White House chief of staff), told the House Armed Services Committee in 2014 that his estimates place revenue closer to $84 billion annually, although his figures include the distribution of cocaine worldwide.[24] Kingpins like Chapo oversee organizations as sophisticated as any multinational—perhaps even more so, given the need to perform the entire process in utter secrecy and often under the risk of arrest or worse. In fact, according to one news profile, the Sinaloa Cartel "might be the most successful criminal enterprise in history."[25]

Whatever the speculated revenue of drug trafficking, one thing is fairly certain—it is highly unlikely that any licit Caribbean endeavor could compete in size, influence, and presence. Today, 97 percent of analyzed U.S.-bound drugs emanate from one country, Colombia, and much of it passes along Caribbean routes.[26] And the trade is crushing local governments. The UN Drug Control Program estimates total regional earnings for the illicit drug industry at almost half the gross domestic product of Jamaica or Trinidad. Suriname, a small South American nation on the southern tier of the Caribbean Basin, has been entirely subsumed as a transshipment point for drug running—no other enterprise in the country exceeds the proceeds that drugs generate, according to one assessment.[27]

The same is true for many of the small island nations and littoral states dotting the Basin. And yet, despite the apparent efficiency with which illicit markets distribute their products, there is a great deal of experimentation and controlled chaos in the flow of narcotics across the Caribbean. Organized networks succeed in their endeavors by disorganizing, fracturing, and federalizing responsibilities within a broader shared infrastructure. The Sinaloa Cartel, for example, is occasionally referred to as the Federation because of its semiautonomous structure. This structure isolates risk and produces more responsive operational models. It also makes it even more difficult to identify basic information, such as the size of an organization. Author Malcolm Beith speculates that Chapo had as many as 150,000 employees at any given time. Georgetown University professor John Bailey, on the other hand, projects the number may be as low as 150.[28] Part of the discrepancy stems from the varied characterization of salaried employees and contractors, a distinction with considerable implications for our conceptualization of what it means to be an organized criminal network. Narcotics trafficking is highly adaptive and creative. Combined with immense capital resources and strong demand in the United States, this creativity engenders myriad routes, tactics, and modes of transportation.

Current Routes and Trends
As expected, after the 1990s boom in Caribbean Sea routes, American anti-smuggling efforts emanating from Florida surged, again opening a resource gap in enforcement along the Mexican border. Cocaine interdicted in the Caribbean in the late 1990s and 2000s dropped to only 5 percent of total intercepted cocaine while the Mexican route flourished.[29] That figure roughly held until 2012, when the percentage of overall cocaine interdicted in the Eastern and Central Caribbean rose to 9 percent.[30] The Mexican drug wars, which began in earnest in late 2006, hastened a process of market reconquest for drug traffickers on trans-Caribbean routes. With law enforcement presence growing on the Mexican border amid the murderous competition of rival Mexican cartels, trafficking was ebbing back toward the Eastern and Central Caribbean once again.

Several anecdotes support this observation although, as with any illicit trade, we are forced to extrapolate heavily from the few visible points of reference. The rise begins seriously in 2013, when U.S. Customs and Border Protection (CBP) reported a 483 percent increase in cocaine washing up along Florida's coast compared to the previous year.[31] Sizeable increases in drug trafficking in Puerto Rico made headlines the following year, spurring speculation that intercartel violence in Mexico had precipitated a shift in the most economical routes. The agent in charge of the

Drug Enforcement Administration's (DEA) Caribbean division suspected these figures would not portend "just a one- or two-year blip."[32] That appears to have been prescient. By 2017 the numbers continued to suggest a resurgent Caribbean pathway. Southern Command reported, in that year's posture statement, that it had recently recorded the highest volume of trans-Caribbean cocaine shipments in a decade. While the Caribbean never ceased to serve as a highway for illicit distribution chains, indicators such as these suggest an invigorated interest in the maritime domain as an avenue for transport.

The *International Narcotics Control Strategy Report* (*INCSR*), a product of the State Department's Bureau of International Narcotics and Law Enforcement Affairs, is the signature U.S. government document on narcotics trafficking. The report identifies two primary categories in the narcotics trade: major illicit drug producers and major drug-transit countries. The report and federal law define a major illicit drug-producing country as "one in which: (A) 1,000 hectares or more of illicit opium poppy is cultivated or harvested during a year; (B) 1,000 hectares or more of illicit coca is cultivated or harvested during a year; or (C) 5,000 hectares or more of illicit cannabis is cultivated or harvested during a year, unless the president determines that such illicit cannabis production does not significantly affect the United States."[33] A major drug-transit country is defined as one "(A) that is a significant direct source of illicit narcotic or psychotropic drugs or other controlled substances significantly affecting the United States; or (B) through which are transported such drugs or substances."[34] The 2017 *INCSR* lists twenty-two countries as either major production or distribution countries of significance. Seventeen are in Latin America, fourteen of which dot the Caribbean Basin: the Bahamas, Belize, Colombia, Costa Rica, the Dominican Republic, Ecuador, Guatemala, Haiti, Honduras, Jamaica, Mexico, Nicaragua, Panama, and Venezuela. Venezuela is further isolated, as one of only three nations to be identified as "having 'failed demonstrably' during the previous twelve months" to address narcotics trafficking and production.[35] The report concludes with a country-by-country assessment of counternarcotics efforts. The State Department's report also identifies major precursor chemical source countries. Precursor chemicals are necessary for the processing of some harvested narcotics and are central to the production of synthetic drugs like methamphetamine or ecstasy. The Caribbean Basin hosts eight major precursor chemical source countries (up from six in the 2014 version of the report): Belize, Colombia, Costa Rica, the Dominican Republic, Guatemala, Honduras, Mexico, and Venezuela. Precursor chemical interdiction is incredibly complicated given the myriad licit uses for many of the products labeled as precursors.

Varying sources and limited data paint an incomplete picture of the most frequented Caribbean trafficking routes. Of the states listed above, commonly noted Central and Eastern Caribbean transshipment points typically include the Bahamas, the Dominican Republic, Haiti, and Puerto Rico, although (as we will see) almost no island is untouched. Puerto Rico was fast becoming a significant destination for traffickers before Hurricane Maria (2017), and James Zackrison, an intelligence analyst at the Office of Naval Intelligence, notes that Trinidad and Tobago, Curacao, Bonaire, Cuba, and a host of small islands in the Lesser Antilles have all been identified as popular transshipment sites.[36] In the Lesser Antilles, many of the English-speaking islands have featured as transshipment points, according to Zackrison, including Antigua and Barbuda, Dominica, Grenada, St. Kitts and Nevis, St. Lucia, St. Vincent and the Grenadines, Anguilla, Monserrat, and the British Virgin Islands. The Greater Antilles, meanwhile, often serve as launching pads for delivering drugs to Puerto Rico, Florida, New York, or other U.S. ports of entry. Marijuana exportation is a smaller business for Caribbean states but is dominated by Jamaica, St. Lucia, and St. Vincent and the Grenadines (according to both Zackrison and the DEA).

With respect to methods of transport, the DEA reports that traffickers in the Bahamas, Haiti, and Puerto Rico consistently use the most diverse means of conveyance, including go-fast boats, containers, airfreight or air drops, and hidden compartments in vessels.[37] Traffickers in Barbados and the Dominican Republic primarily use go-fast boats and containers while those in Trinidad and Tobago appear to predominantly prefer go-fast boats. A Panamanian-flagged vessel interdicted by the U.S. Coast Guard 780 nautical miles west of Peru in 1995 demonstrates just how internationalized narcotics trafficking is; documentation on board confirmed it had smuggled cocaine all across the Caribbean.[38] The ship, *Nataly*, called in Belize, Colombia, the Dominican Republic, Haiti, Honduras, Panama, Puerto Rico, Suriname, Trinidad and Tobago, Turks and Caicos, and even Louisiana. The *Nataly*, eventually intercepted in the Pacific Ocean, had also called in Ecuador, demonstrating the relative simplicity of moving between oceans.

Given the clandestine nature of smuggling, anecdotal information from ships like the *Nataly* offers some of the best insight into an otherwise opaque trade. Often traffickers cross the transit zone undetected and unmolested, navigating across crowded littorals, past small fishing villages, and into the open ocean to evade detection. In the cross-border effort to remain under the radar, both literally and metaphorically, traffickers are aided by the diversity of political and geographical climes available within such a small radius. Each island and littoral state provides a unique set of geopolitical factors, enabling trafficking in a variety of ways.

Trinidad and Tobago, for example, is only seven miles from mainland South America and serves as an ideal route for traffickers leaving Venezuela via the Orinoco River—police are prevented from pursuing them beyond the international border.[39] The proximity also means that small local craft can make the journey quickly and inconspicuously. St. Vincent and the Grenadines is similarly well-positioned geographically and offers an added allure with lax flag of convenience ship registration that has attracted Colombian smugglers for decades.[40] Flag of convenience laws allow ship owners to register vessels in states other than their own, which in turn makes it difficult to implicate ship owners in any crime undertaken with the use of the vessel. The Bahamas, to the north, provides a different set of attracting qualities. Less than one hundred miles from Florida, the Bahamas offers hundreds of islands and keys within a stone's throw of the mainland United States. It also sits astride major sea lines of communication, serving as an interchange between Panama Canal traffic, shipping passing throughout the Atlantic Ocean, and the Gulf of Mexico's major ports and oil terminals. All of this is contained within a national oceanic jurisdiction of more than 250,000 miles—far greater than the Royal Bahamas Defense Force can reasonably patrol.[41]

The most significant conclusion from an overview of narcotics trafficking in the Caribbean is the versatility with which suppliers adapt to meet the pressures of demand and enforcement. In the maritime domain, "everything from modern container ships to rusty small freighters, to fishing boats, to small 'go-fast' craft" have been used to move drugs.[42] Frequently, trafficking through such means, especially small boats, is cheap and unsophisticated. Go-fast boats are a popular means of conveyance because, other than GPS and cell phones, they are largely free of complex instrumentation. Yet, just as narcotics traffickers have expanded their geographical routes and methods, so too have they experimented with the physical means of conveyance. One of the fastest-growing alternatives to the ubiquitous Caribbean go-fast boat is the narco-submarine (or narco-sub). A narco-sub is a "custom-made, self-propelled" submersible or semisubmersible used to transport narcotics and related goods.[43] The Colombian cartels in particular have featured heavily over the past two decades in the development of submersible technology, even buying at least one submarine from the Russian mafia.[44] It is within the last decade, however, that narco-subs have gained serious traction in response to repeated successes in narcotics interdictions in the mid-2000s. In part, go-fast boats began carrying smaller loads across more craft to mitigate risk. And increasingly, narco-submarines began to play an important role in the effort to diversify maritime smuggling techniques.

Narco-submarines come in two varieties, self-propelled semisubmersibles (SPSS) and fully submersible vessels (FSV), as described in testimony by former Joint Interagency Task Force South (JIATF-South) commander Rear Adm. Charles Michel (USCG). They are purpose-built machines, designed by engineers, constructed in parts under the thick cover of the Amazon forest, and floated like logs downriver for assembly and deployment.[45] JIATF-South first detected an SPSS in operation in 2006. In less than three years, the task force had identified up to sixty SPSS events annually, delivering more than 330 metric tons of cocaine. In the first six years of the submarines' use, JIATF-South confirmed 214 documented SPSS events, although only forty-five were disrupted. The Colombian navy has also facilitated the interdiction of several diesel-powered submersibles, each carrying a similarly impressive cache of cocaine.[46] At a cost of approximately $1 million a vessel, each narco-sub has the capacity to reap one hundred times that figure in revenue. For that reason, such craft are expendable for traffickers. In the unlikely event of interdiction, the submarine can be flooded with water, taking evidence to the depths while operators are pulled from the sea.[47] The fully submersible variant can traffic a hefty ten metric tons of cocaine up to 6,800 nautical miles, according to testimony by General Kelly. That is enough range to reach as far as Africa. From Colombia's Caribbean littoral, Kelly notes, an FSV can venture to Florida, Texas, or California in ten to twelve days, likely without a hint of detection.[48] As one of many methods of distribution, narco-submarines provide a resiliency and carrying capacity that have quickly filled an important niche in bulk transnational crime. While far fewer in number than go-fast boats, SPSSs and FSVs potentially pose "an even more insidious threat to the security of the United States," in the words of Rear Admiral Michel.[49] Such narco-subs exemplify the nature of the hybrid threat. In many instances, the actors in control of such vessels retain criminal motives. Yet the construction and deployment of these craft (not to mention the cultivation of the products they carry) frequently require political control of territory. And their capacity to move not just drugs but people, money, and weapons (all of which I explore in the next two chapters) means that such submersibles encapsulate an inherent capacity to threaten international security.

Trafficking Flows
By the mid-2010s the new jewel in the Caribbean trafficking crown was Puerto Rico. Within days in April 2014, investigators on the island of Puerto Rico intercepted two boats, each with millions of dollars of cocaine stashed on board. In the first instance, law enforcement uncovered 1,774 kilograms of cocaine; in the second, agents uncovered 1,530 kilograms. In

May 2014 the *New York Times* published a profile on narcotics smuggling on the island.[50] The reason for this rise, and the interest in it, is readily apparent; as Kelly notes, "Once cocaine successfully reaches Puerto Rico, it has reached the U.S. homeland."[51] Yet Puerto Rico is only one stop on a long journey. Traditionally, the island of Hispaniola has served as a midway point in this trip. The year before the *New York Times* profile, U.S. Southern Command noted an increase of 3 percent (to thirty-two metric tons) of known cocaine passing through the island. Haiti, on the western end of Hispaniola, would be the obvious locus of these shipments, plagued as it is by mismanagement and natural disaster. Yet it is the island's larger eastern state, the Dominican Republic, that may be the Caribbean's largest island transshipment hub. As the "first point of contact for Puerto Rican operators," the Dominican Republic plays a central role in an axis passing up through Puerto Rico, Miami, and New York.[52]

Dominicans make up a considerable portion of the immigrant population on the East Coast of the United States, particularly in New York, so much so that the sociologist Anthony Maingot considers Dominicans to have forged a fully "binational" society split between the United States and the Caribbean.[53] Better standards of living in the United States also mean that New York–based Dominicans contribute heavily to the Dominican economy through remittances. As David Kilcullen explores in *Out of the Mountains*, remittance networks are one of the many ways that communities remain interconnected (a "megatrend") with the wider world. Kilcullen demonstrates this effect through Jamaican posses, which extort Jamaica's massive remittance system to fund illicit ventures and their rule of garrison districts. International ties like remittances also contribute to trafficking, concealing illicit flows embedded within licit traffic. In much the same way, the heavy exchange of remittances, people, and goods between Dominican binational communities forms the basis for many Dominican smuggling routes.

As I explain in the following chapters, this fundamental infrastructure of smuggling is a commodity in its own right. Such networks are by nature highly adaptable and can be easily co-opted by other actors or for the transport of alternative products. Thus, the existence of Dominican smugglers operating between the Caribbean and the United States has attracted attention from more than just law enforcement. Maingot relays a DEA assessment that Colombian actors were directly coordinating Dominican networks and routes in an effort to capitalize on this preexisting infrastructure. Whether organized by external forces or internally connected by ethnic ties, the centrality of the Dominican–Puerto Rican drug route now pervades Dominican language and culture. Dominican immigrants living in the United States are sometimes called "Dominicanyorks," as explained by Maingot, and have faced hostility back in the Dominican Republic "in major part because they have been stereotyped as being drug dealers."[54]

Trends rarely persist in the cat-and-mouse game of narcotics smuggling. As Puerto Rico grows as an attractive destination for Caribbean smuggling operations, efforts to evade detection and simplify shipment routes have resulted in increased attempts to bypass Hispaniola entirely.[55] According to a Puerto Rico–based Immigration and Customs Enforcement agent, "Cocaine is increasingly trafficked in larger amounts directly north from Venezuela on boats that refuel at sea during the roughly two-day voyage."[56] It is still too early to tell how Puerto Rico's devastation in 2017 at the hands of Hurricane Maria might alter this pattern over the long term. Moreover, with the stronghold the Dominican Republic has on transshipment, and the strategic placement of ethnic and familial ties in New York, it is unlikely that Hispaniola will disappear from the trade. Nevertheless, as history proves, the drug market is highly adaptable. While squeezing the balloon at one end (i.e., the Dominican Republic) may decrease cocaine traffic through the island, experience suggests that the excess "air" will simply find a place elsewhere.

One such "elsewhere" is Jamaica, which originally made its mark in drug trafficking through marijuana. Zackrison notes that Jamaica, alongside Belize (and a few smaller islands), grows and exports marijuana for regional consumption, mostly in the Bahamas, Guyana, Trinidad and Tobago, and the Eastern Caribbean. Inevitably, with the dramatic rise in sale price to American consumers, some Jamaican marijuana eventually made its way into the United States as well. Jamaica remains the largest Caribbean supplier of marijuana to the United States according to Mary Alice Young, a law professor at the University of the West of England.[57] Yet, as we noted with Dominican smuggling networks, the infrastructure of trafficking is easily co-opted, and illicit marijuana distribution provided an avenue for more expansive smuggling operations. A similar trajectory took place in St. Kitts and Nevis. There marijuana production attracted smugglers who sought to co-opt "the clandestine links already established by the locals" to develop alternative cocaine trafficking routes.[58] Some Jamaicans, especially those who make their living on the coast, are made uneasy by the prospect of similar infiltration migrating into their towns. One news story notes the "anxiety" of some residents of the coastal Jamaican hamlet of Forum, whose fishermen worry about an encroaching narcotics business in their quiet village. And in much of Jamaica, that anxiety has long since been validated. The central Caribbean region has reemerged as a route of favor for traffickers, and Jamaica appears to play a significant role in that nexus.

The scene along parts of Jamaica's southern coast already threatens hamlets like Forum, where coastal police embarked in speedboats and armed with assault rifles now dot the shores.[59]

In the first half of 2013, cocaine seizures in Jamaica doubled over the previous year.[60] Yet that doubled figure only came to 354 kilograms, paltry when compared to individual hauls of more than 1,000 kilograms in Puerto Rico and elsewhere. When, in April 2016, Jamaican authorities staged the largest bust in more than a decade, the estimated street value was still on the low end—about $7 million.[61] Thus, while Jamaica is widely suspected of being an increasingly popular transshipment point for northward bound cocaine, it does not rank highly on the list of Caribbean seizures by volume. The U.S. government (in the *INCSR*) nevertheless maintains that Jamaica is a "major drug-transit country."[62]

While interdiction rates lag, tight-knit island communities exhibit prescient concerns that, as seen in chapter 2, can have far-reaching consequences for public safety. The corrosive impact of the trade has made itself felt throughout the social fabric of Jamaica, as expressed by the fishermen of Forum. The national security minister, in a speech to the Jamaican parliament a decade earlier in 2002, called the narcotics trade a "taproot" for further instability and violence.[63] While poor interdiction numbers offer little empirical support for claims of Jamaica's rise in the drug trade, sensitive popular sentiments such as these do.

Drug abuse serves as yet another indirect indicator of drug trafficking and only serves to fuel social tension. Payment in kind, instead of in cash, is a common financing alternative for drug networks. This invariably breeds drug addiction growth within communities, and Jamaica is one such victim of this phenomenon. The country now "harbors thriving cocaine and marijuana markets," according to Young, which are still further evidence of the island's heavy use as a transshipment point.[64] Yet, while the Eastern and Central Caribbean may be growing in popularity among criminal organizations, these routes still pale in comparison to the sheer size of the trade along the Western Caribbean littoral.

According to Rear Admiral Michel's testimony, "The Mexico/Central American corridor, which includes the waters of the Eastern Pacific and the Western Caribbean, is the primary threat vector toward the United States, accounting for more than 90 percent of total documented cocaine movement." And 80 percent of this traffic makes first landfall in Central America, with the remainder heading directly for Mexico, to the best knowledge of the State Department. About 35 percent of all outward-bound cocaine makes landfall in Honduras alone, according to Michel. Moving primarily in go-fast boats—"open hulled boats anywhere from 20 to 50 feet in length with 1 to 4 powerful outboard engines"—traffickers move as much as 3.5 metric tons of cocaine as they ferry between Honduras (principally) and Colombia.[65] Of the hundreds of go-fast events documented by JIATF-South's Consolidated Counter Drug Database, nearly all were detected in Central American littorals alongside Honduras and Guatemala, then up to Mexico. Increasingly, narco-submarines are also in use, and the DEA estimates as much as 30 percent of the maritime movement of drugs in this corridor is now undertaken by these vessels.[66]

This traffic, between South America and Central America, is the primary instance of maritime conveyance for most cocaine exiting the source zone. Here, between Central America and Colombia and Venezuela, 80 percent of U.S.-bound cocaine first moves by water.[67] Once in Central America, much of the onward smuggling takes place over land. Of those shipments that do progress on water, many embark along the Western Caribbean shoreline, using the Bay Islands as a point of departure.[68] The Bay Islands are within reach of the small Colombian islands of San Andrés and Providencia, and many inhabitants trace their lineage to English-speaking immigrants from Belize, Jamaica, and other British Caribbean possessions. Thus, Bay Islanders frequently share ethnic and economic similarities with English-speaking Caribbean communities across the Basin as well as a shared history of "small-scale smuggling," according to Maingot. Its geographical position at the crossroads of Central America, South America, the Western Caribbean, and Mexico, alongside historical and linguistic ties to varying parts of the Basin, all converge to make Honduras and its islands a predictable victim of the narcotics trade.

The United States regards all seven of the Central American states—Belize, Costa Rica, El Salvador, Guatemala, Honduras, Nicaragua, and Panama—as "major drug transit countries." And while they persist as actors in the trade, their individual roles are constantly in flux. As traffickers diversify their product lines to include designer drugs like methamphetamine, the littoral states of Central America are now playing a greater role in the trafficking of precursor chemicals. Multiple tons have been seized in Belize, El Salvador, Guatemala, and Honduras. Given the lack of climate and soil requirements for the production of synthetic drugs, the passage of precursor chemicals through Central America and the Caribbean may portend a geographical expansion of methamphetamine and ecstasy production.[69]

Counternarcotics

It is easy to assume that combating narcotics trafficking should not require a theoretically grounded, multidimensional strategy of maritime security.

Surely stopping enough of the supply from reaching U.S. shores would change the cost calculus for traffickers. This thinking has characterized a large portion of American maritime enforcement efforts, like Operation Martillo, which "seeks to deny the use of the Central American littorals by TCOs [transnational criminal organizations] ... while maximizing the drug interdiction efforts of our interagency partners."[70] By increasing enforcement in the littorals, Operation Martillo attempts to leverage the balloon effect to U.S. authorities' advantage, forcing smugglers back into vulnerable deep water where coast guards and navies have the advantage. As Coast Guard commandant Adm. Paul F. Zukunft noted, traffickers have "very few allies at sea, and that's where we do have the upper hand."[71] Yet, as I discuss below, interdiction-based enforcement efforts are generally inefficient and potentially even fatally flawed. (This interdiction model is in the same mold as the midcentury American policing model discussed in chapter 2. This strategy of policing focuses on quick responses and catching crimes in the act. Yet, in both policing and drug interdiction, officers frequently cannot respond quickly enough, and to enough incidents, to reach a tipping point.)

This is not to say that Martillo and operations like it have not seen success. In the operation's first two and a half years, as General Kelly noted in testimony, Martillo disrupted a total of 272 metric tons of cocaine, seized $10.7 million in bulk cash, and seized 198 vessels and aircraft. JIATF-South, which coordinates Martillo under Southern Command, reported declines in illicit maritime activity in Western Caribbean littorals of 43 percent in those first years. There is no broader evidence, however, that this has precipitated a decline in overall trafficking. The resilience of the narcotics industry is astonishing given that Southern Command commander Adm. Kurt Tidd estimated in 2017 testimony that the command had disrupted almost $5.5 billion in cocaine trafficking the previous year.[72] As the ceaseless back-and-forth continued, Southern Command had to wonder if success closing direct routes to Honduras would propel traffickers back again into the littorals.[73] Without a massive increase in resourcing, this balloon effect will continue to dictate ambiguous enforcement outcomes. Limited assets, deployed in limited locations, will inevitably create enforcement gaps elsewhere.

And those resources are not likely to come. One Southern Command commander told Congress (speaking of drug trafficking), "I simply sit and watch it go by."[74] Sequestration-induced cuts made particularly deep incisions into Southern Command's capacity, which already sees itself as "an economy of force Combatant Command," in the words of General Kelly.[75] Southern Command, as expressed by Kelly, perceives itself "as the lowest priority Geographic Combatant Command" and expected to receive "little, if any, 'trickle down' of restored funding."[76] He further testified that the forces under his command failed to capture an astounding 80 percent of estimated Caribbean-trafficked narcotics. Little has changed. Kelly's successor, Admiral Tidd, noted to Congress that JIATF-South was still unable to interdict three-quarters of known events because of shortages in resourcing.[77] And in Kelly's assessment, Southern Command only receives about 5 percent of the resources necessary to pursue this mission.[78]

In an effort to offset cuts in resources, Kelly relied "heavily on the U.S. Coast Guard and Customs and Border Protection, which now provide the bulk of the ships and aircraft available."[79] Despite the Coast Guard's tonnage compared to most navies, however, the force still receives only a fraction of the resources it would likely need to interdict enough narcotics to stem the flow. Resourcing is an issue even in the Coast Guard's primary areas of operation, as sequestration further imposed budget tensions and threats of downsizing. Admiral Michel said in 2012 that his Task Force was "unable to target 74 percent of high confidence events,"[80] in line with General Kelly's testimony (2014) and that of Admiral Tidd (2017), because of resource scarcity. Such financial constraints are not strictly an expression of Congress' frugality, either. To put this resource deficiency to scale, the 2017 *INCSR* notes that in one year the Coast Guard employed "2,200 cutter days, 1,400 Airborne Use of Force capable helicopters days, and 3,100 surveillance aircraft hours on counterdrug patrols."[81] Added to that, Kelly notes that U.S. Naval Forces Southern Command provided a total of "six frigates, [one] High Speed Vessel (HSV) SWIFT, four fixed-wing Maritime Patrol aircraft and two Scientific Development Squadron ONE detachments" in support of Operation Martillo.[82] Still, these resources are woefully insufficient. Even under normal budgetary circumstances, the maritime enforcement bodies responsible for interdiction in the Caribbean will at no point in the foreseeable future command the expansive resources necessary to combat narcotics trafficking through interdiction-led initiatives. The task is simply Sisyphean.

Not to oversimplify matters, there is a great deal of experimentation in attempts to better counter narcotics trafficking. U.S. Marine Corps Forces South, as noted by Kelly (a Marine himself), leads Joint Riverine Training Teams to train local littoral and riverine police, coast guard, and navy units in best practices for interoperating between land and sea. Operation Unified Resolve, launched in September 2012, promoted interagency Puerto Rican interdiction efforts across JIATF-South, Coast Guard District Seven, CBP, and others.[83] In late 2013 Southern Command launched a collaborative counterthreat finance branch with the Department of the

Treasury to map and combat illicit networks through financial means. In many respects, these operations are a model for "whole of government" efforts. U.S. Southern Command headquarters contains thirty-four representatives from fifteen distinct federal agencies on its staff.[84] Kelly maintained that the "maturity, strategic leadership, and tactical collaboration between JIATFS, the USCG, and CBP have greatly enhanced the effectiveness of countering illicit trafficking (CIT) operations."[85] Pointedly, the Navy's ability to integrate with law enforcement partners and to support a predominantly constabulary mission lends credence to the notion that the service can and should engage more substantively with maritime security issues.

International collaborations also play a significant role in Caribbean operations, both as a means of partnership and capacity building but also in an effort to leverage limited resources. Southern Command reported under Kelly's tenure that, of the nearly 150 interdictions supported by JIATF-South in a given year, half would have failed without some degree of international support. Rear Admiral Michel told Congress that international partners were active participants in 83 percent of disruptions the preceding year, acting as a "force multiplier." Alongside the aid of Caribbean enforcement agents, larger international players also assist in counterdrug efforts. Some of the most significant of these partners include those with regional dependencies, territories, or associated states such as the United Kingdom, France, and the Netherlands, as well as Canada.[86]

These partnerships provide an answer to an important issue in maritime security, the Posse Comitatus Act, the regulation forbidding the U.S. armed forces from acting in a law enforcement capacity. Southern Command has directly addressed this by embarking Coast Guard law enforcement detachments on board naval ships. The Coast Guard operates in both a military and law enforcement capacity and is exempt from Posse Comitatus. As reported in the 2017 *INCSR*, the U.S. Coast Guard routinely deploys law enforcement detachments not only on board U.S. Navy ships but also on British, Dutch, and Canadian ships. RFA *Wave Knight* of the Royal Fleet Auxiliary, for example, supported multiple interdictions with the aid of one such team, even hosting a U.S. Coast Guard helicopter for more than half of the ship's 471-day deployment—the first time ever that a tactical Coast Guard helicopter squadron had deployed on a non-U.S. ship.[87] In the span of only two months in early 2014, RFA *Wave Knight* made or facilitated at least three interdictions, one seizing drugs worth up to £300 million according to news reports, with the facilitation of the U.S. Coast Guard. Other multinational operations have included a wide array of partners, like one Interpol-led investigation that coordinated thirty-four countries, Europol, and several constabulary forces such as the French Coast Guard and the Royal Canadian Mounted Police.[88] International partnerships in the Caribbean, while often threat-specific (which, as I argue in the ensuing chapters, is a suboptimal approach) are widespread and emblematic of the structures needed to fully address transnational issues.

Mutual legal assistance treaties are another important tool, both legally and strategically, in international partnerships. Similar arrangements date as far back as 1926, when the British government allowed the U.S. Coast Guard to interdict ships in the Bahamas suspected of smuggling alcohol into the United States.[89] Interdiction arrangements have since continued to play an important role in the Caribbean, where mutual legal assistance treaties facilitate the exchange of information and coordinate international law enforcement efforts. The United States maintains such agreements with at least seventeen governments covering at least twenty-six countries and territories.

The United States has also invested heavily in such partnerships. Southern Command, as testified to by Kelly, has provided millions of dollars in counternarcotics training and equipment in recent years, including delivering forty-two interceptor boats to regional navies and coast guards. The United States has also deployed technical assistance field teams to an expanding number of Caribbean maritime forces to offer logistical and mechanical training.[90] In these and similar programs, the Navy, Coast Guard, and Marine Corps have demonstrated an inclination toward creative problem solving, which together suggests that the Caribbean is primed to serve as a proof of concept (perhaps even a testing ground) for new strategies of maritime security.

Lessons for Maritime Security

As dramatized in the case of narcotics trafficking, the capacity to transfer an illicit product between jurisdictions is a commodity in its own right. In the words of former *Foreign Policy* editor Moisés Naím, "borders are a trafficker's best friend," posing a serious obstacle for law enforcement. And while "this asymmetry is not new," Naím concludes that it has "become more acute."[91] The infrastructure and networks required to facilitate this kind of movement are immense, clandestine in character, and by nature resistant to the efforts of single actors. Even the myriad international partnerships in the Caribbean struggle to overcome the fundamental obstacle of international borders. Yet the single greatest impediment to combating narcotics trafficking is the perception that simply catching enough "bad guys" or interdicting enough cocaine will solve the problem.

Even in instances of high confidence intelligence, the commandant of the Coast Guard noted, "There aren't enough ships to catch them all ... We're giving 60 percent of what we know, literally, a free pass."[92] This is all the more noteworthy considering the war on drugs represents "the largest deployment of money, technology, and personnel that humankind has ever devoted to stopping drugs from moving across borders," costing the United States an estimated $200 billion annually in direct health care and enforcement costs related to cocaine alone.[93] Given the scale of the narcotics trade (Zackrison relays a DEA estimate that $60 billion in illicit proceeds are laundered through the Caribbean annually[94]), even frequent high-quantity interdictions have done little to slow the movement of drugs worldwide. All this suggests that the issue of narcotics trafficking is not one to be solved by sheer scale of investment. The value of drugs entering the United States is simply too high for the United States to impart a meaningful enough penalty to deter trafficking.

Since the launch of the war on drugs, the United States has constructed an immensely expensive apparatus devoted to counterdrug operations. At the federal level alone, the DEA, Immigration and Customs Enforcement, CBP, the Marshals Service, the Secret Service, the Federal Bureau of Investigation, the Coast Guard, and countless other national agencies are funded for counterdrug missions at an estimated price of $20 billion annually, according to Naím.[95] Yet, despite the colossal size of the counterdrug apparatus, the task of detection and interdiction is exponentially larger.

To be sure, there have been several successes over the years. Aid money to Colombia helped decapitate the leadership of the Medellín and Cali cartels, which helped dismantle large portions of their organizations. Weekly news reports relay daring arrests and high-profile corruption charges. Such successes come with little reward, however. Even when transnational organized crime was structured in a more hierarchical format, Naím relays that the death of a kingpin like Pablo Escobar had virtually no impact on cocaine production and distribution. Later, he notes, arrests of top leadership of both the Ramón Arellano Felíx criminal organization and the Gulf Cartel in the early 2000s were similarly met by imperceptible changes in gross narcotics trafficking. More significantly, decapitation of cartels and a focus on interdictions ignore important changes in the structure of trafficking. As I explored earlier, Colombian organizations began losing dominance when they enabled Mexican transnational criminal organizations to act as wholesalers. Today that process has continued to devolve the narcotics trade. Crossing borders adds the greatest value to cocaine as it moves, shifting power and earnings to the "middle of the distribution chain, to the points where the greatest opportunities exist for high-value cross-border transactions, diversification, and strategic partnerships."[96] Opportunistic middlemen constitute networks of shared smuggling infrastructure moving goods in shifting patterns shaped by the demands of the market. The arrest of a distant kingpin in Mexico, or the interdiction of a small shipment in Jamaica, means relatively little to a facilitator in the Dominican Republic or Honduras.

It is not surprising, therefore, that counternarcotics operations have met limited success, despite being the signature American maritime security task in the Caribbean. As we have seen, the reasons for this failure are born out of two significant misperceptions. First, as Naím puts it, "We still speak of drug 'cartels,' but the drug business today has largely dissolved the heavy organized crime-like operations of the past and works in more nimble, less traceable ways."[97] The perception of transnational organized crime as corporately hierarchical instead of fluidly franchised produces flawed strategies, most notably institutional decapitation. Hybrid threats in the modern age are frequently characterized by decentralized actors operating in formations more akin to swarms than coordinated groups. Decapitation of senior leadership makes little impact on the semiautonomous middlemen and small-scale actors who carry out the drug trade. In the Dominican Republic, for example, Maingot depicts the resilience of organizations with no central node. In the absence of a cartel with a select few lieutenants and capos at the top, he writes that "the state fights a Hydra of decentralised and localised gangs without hierarchy or enduring central control."[98] Second, narcotics trafficking is not a conventional security issue. It is simultaneously too extensive for the Coast Guard to handle alone and too unconventional to arouse much enthusiasm from the Navy. JIATF-South's mission is to deny traffickers the use of more than six million square miles of sea and airspace, according to the Coast Guard. Despite accounting for more than half of all drug seizures by U.S. authorities, the Coast Guard (alongside the Navy, CBP, DEA, and others in the Task Force) is fighting a losing battle.[99] The answer, I propose, comes not just from budgets or force structures but more fundamentally from how we frame the problem.

Sailors assigned to Coastal Riverine Group 1 play board games on board expeditionary fast transport USNS *Spearhead* (T-EPF 1) during Southern Partnership Station 2017 (SPS-EPF 17). SPS-EPF 17 is a U.S. Navy deployment executed by U.S. Naval Forces Southern Command/U.S. Fourth Fleet, focused on subject matter expert exchanges with partner nation militaries and security forces in Central and South America. *U.S. Navy photo by Mass Communication Specialist 2nd Class Brittney Cannady/Released*

CHAPTER FOUR
Trafficking
Guns, People, and Terror

GUNS

Narcotics trafficking garners international concern because it crosses borders. Yet the individual criminal incidents that collectively constitute transnational crime necessarily occur within nations, districts, and communities. The adage "all politics is local" is equally true of transnational crime. Even global enterprises must operate within local contexts. In the Caribbean Basin, that context is often informed by alarming crime rates, including some of the highest murder rates in the world. Local crime is therefore the dominant political concern for many of the region's inhabitants. And because local crime and related social grievances monopolize the limited resources of Caribbean lawmakers, regional security is typically left to foreign bodies, financed (and, as a consequence, tailored) to the needs and assumptions of someone else's strategic perspective.[1] Engaging and incentivizing the necessary participation of Caribbean partners to combat illicit trafficking requires incorporating the region's emphasis on the home front.

The scale and scope of narcotics trafficking would imply that quite a large number of individuals participate in or facilitate the movement of illegal substances. In fact, organized or semiorganized networks cannot keep pace with demand. To keep product moving across the American border, large criminal groups frequently license the use of their routes to small or individual actors. Despite "taxes" of up to 60 percent, moving narcotics is so lucrative that there is no shortage of part-time traffickers to fill demand.[2] Such was the case with a Mexican business owner, relayed by Moisés Naím, ascribed the pseudonym Don Alfonzo. Alfonzo uncovered that some of his employees were using company vehicles to transport narcotics to supplement their incomes (one successful trip was worth a year's salary). Instead of turning them into the police, Alfonzo approached them with a proposition. He would finance their purchase of narcotics (the largest up-front cost) and allow them to continue using his vehicles to cross the border while taking a percentage off the top. Alfonzo's successful small-scale network operates largely independent of the major Mexican cartels. Over the years, businesses like this have done well, largely existing outside both intercartel violence as well as the multi-billion-dollar international counternarcotic effort.[3] Such entrepreneurs are therefore aware that drugs offer the prospect of significantly enhanced supplemental income at minimal risk—Alfonzo jokes, "I think in this town the construction business is riskier than moving drugs." In Alfonzo's words, "The gringos and the police here are busy chasing the big guys, and if they go after the small guys like us they will need to build a new jail the size of this whole town.... No government can touch this. Why should they? The big guys give good theater and are good for politics. We don't."[4]

Even many of those who do not physically transport drugs—by choice, in Alfonzo's case, or by force, in many others—are implicated in the trade. The logistical challenges of moving drugs across vast distances require an extensive support network. Costa Rican coast guard director Martín Arias describes how a boat laden with cocaine traveling from Colombia to Mexico "could require as many as 24 other boats along the way to successfully complete the journey."[5] David Chacón, the president of a fishing cooperative on Costa Rica's Pacific coast, explains that local fishermen are press-ganged by traffickers to rendezvous at sea with fuel and rations as part of a series of way stations. In Costa Rica, as with many Caribbean and littoral Central American states, the existing coast guard forces are ill-suited to police the entirety of the country's borders, leaving exposed the "economic vulnerability" of small fishing communities to predation.[6] And there is no shortage of avenues through which communities are enveloped into the trade. Falcons, lookouts employed by the cartels, are paid as much as one hundred dollars a month to keep a weather eye on the pace of border enforcement or police activity and make the occasional phone call. Trafficking is so extensive, and its culture so pervasive, that the Drug Enforcement Administration (DEA) speculates (somewhat hyperbolically, perhaps) that there are cities in which "virtually every cabdriver" moonlights as a watcher for the cartels.[7] The drug culture in the Caribbean is not one that can be addressed only by high-profile arrests or interdictions. Drug abuse and widespread participation in narcotics trafficking erode the efficacy of the state and the self-efficacy of communities.

This permeation of the narcotics business into Caribbean society has an insidious impact throughout many communities. On the low end, police estimate that roughly 60 percent of crime in Caribbean Basin countries is related to drugs.[8] That figure is considerably higher regarding violent, gun-related crimes. In 2004 the Colombian president, Alvaro Uribe, alleged that "more than 90 percent of the crime in Colombia" was perpetrated with illegal firearms.[9] Indeed, perhaps the most significant connection between narcotics trafficking and local crime is its impact on homicides. The reason is simple; much of the revenue from selling drugs is funneled directly into arms trafficking. Just as Jamaica's or St. Kitts and Nevis' marijuana smuggling networks were co-opted for cocaine trafficking—a phenomenon we saw in the previous chapter—so too are cocaine networks co-opted for returning guns to transit and source countries. One source of evidence for this mixed use of trafficking infrastructure is that arms traffickers and narcotics traffickers are often the same groups. As one author notes, there is a deeply intersectional relationship between narcotics and weapons smuggling in Latin America: "Groups involved in narcotics also are the leading traffickers in armaments and much of the money used for arms purchases is generated by the narcotics trade."[10] Once weapons are acquired, they are then used to further entrench the broader trafficking infrastructure by defending transit routes from competition, terrorizing the public into submission, and battling security forces. The avalanche of small arms and light weapons pouring into the Basin has helped it become the most deadly region in the world. Eight of the ten countries with the world's highest homicide rates are in the Caribbean Basin, including Honduras, Belize, El Salvador, Guatemala, Jamaica, St. Kitts and Nevis, Venezuela, and Colombia. And the impact of murder rates that rival those of war zones cannot be understated.

A 1996 survey in Latin America, noted by Naím, identified gun crime as the leading socioeconomic issue facing the region. In the intervening two decades, little headway has been made. In the last ten years, Jamaica has only fallen from number one to number six on the list; the country still experiences more than one thousand murders a year on average. The island has been swamped by the epidemic. Drugs are traded for guns in an exchange that exacerbates both local crime and international narcotics trafficking. The U.S. State Department estimates homicide conviction rates at only 5 percent.[11] According to Omar Davies, former Jamaican minister of transport, works and housing, and former minister of finance and planning (as relayed by Young), "'Vicious criminal organisations' permeate every level of society . . . [and] the loss of lives from serious crime is the biggest burden to the country."[12] Jamaica even struggles to find enough places to house all the bodies. Jamaica has been without a national morgue since the 1970s, according to an Associated Press exposé, resulting in inadequate autopsy facilities and the frequent contamination of evidence.

Bodies are stored in funeral homes that are paid so little (six to ten dollars per body, per day) that they often cannot afford to keep them refrigerated. One consultant to the Jamaican police noted that he has "never seen such deplorable conditions and lack of support from a government to resolve these issues." The former mayor of Kingston, Desmond McKenzie, noted plainly, "The simple fact is the government has failed miserably."[13] Puerto Rico, too, is not immune. In 2012 homicide rates on the island peaked at 26.5 per 100,000 people. This, too, is directly related to narcotics trafficking; the U.S. Coast Guard estimates that 75 percent of homicides in Puerto Rico can now be tied to the trade.[14] Yet, of all the death and gun use in the region, Honduras lays claim to the title of the most violent place on the planet. Much of this violence takes place in San Pedro Sula (the deadliest city on Earth) and along the Guatemalan–Honduran border, both major transshipment chokepoints and the subject of a small-arms survey report.[15]

In Honduras, *transportistas* (trafficking groups) and territorial gangs are in constant struggle over control of the world's most frequented cocaine passageways. Murder rates are highest in these "highly concentrated urban centres and strategic points (the Atlantic coast and border regions)."[16] *Maras* (gangs) and *pandillas* (youth gangs) emerged in Honduras emulating the model of American gangs in the 1980s and 1990s and have been a major feature of these violent turf wars ever since, according to the small-arms survey report. Gang involvement is so high that estimates project there are as many as 151 *maras* members per 100,000 Hondurans. Regionally, the numbers are no better. The report estimates that Guatemala stands at 146 per 100,000, and El Salvador is at a staggering 318 per 100,000 Hondurans. Over about a decade, more than 40,000 people were killed in Honduras, the vast majority with a gun. In 2012 alone, the authors note, Honduras witnessed more than 7,000 violent deaths (about a 200 percent rise in less than ten years), though in later years the numbers were somewhat lower (about 5,100 in 2016, according to media reports).[17] A homicide rate of 85.5 per 100,000 people is more than eight and a half times the global average.[18] Illegal weapons, purchased with the proceeds of drug running and smuggled through many of the same mechanisms, drive Honduras' murder epidemic. As is the nature of illicit trades, estimates on the number of small arms and light weapons in the country are variable. Extrapolating the number of firearms based on the number of homicides would suggest, per the small-arms survey report, the presence of around 420,000 illegal weapons in Honduras. The authors also relay an estimate from the UN Development Programme, which suspects that closer to one-third of the firearms in Central America are in Honduras, placing the number much higher, between 800,000 and 1 million weapons. While the relationship between traffickers, gangs, and firearms may be complex, the hyperbolic case of Honduras demonstrates just how interconnected they are.

And gun violence is not simply at issue because of the rampant homicides (as if that were not enough). Just as with the movement of drugs, the trafficking of guns is itself a lucrative and complex international endeavor. The spike in crime across the Caribbean Basin can be directly tied to the impact of "a combination of drugs and money in the form of the naked display of power, through the use of arms."[19] In 2000, Trinidad's attorney general conveyed the scope of the relationship between transnational crime and local communities, noting that there is a "direct nexus" between the narcotics trade and a range of "other evils," from violent crime to domestic violence. He went on to note that the resultant disorder "changes democratic institutions, erodes the rule of law, and destroys civic order with impunity."[20] The relationships expressed here, between crime, disorder, and perceptions of eroding institutions, are all elements of the broken windows theory as explored in chapter 2. I turn next to making these links even clearer.

The interplay of local crime with international trafficking is not lost in American maritime strategy, if perhaps only combated on a small scale. In testimony to the House Armed Services Committee, Southern Command commander Gen. John Kelly recognized that this "skyrocketing criminal violence exacerbates existing challenges." In his words, the Caribbean Basin (and Latin America) "remains the most unequal and insecure region in the world."[21] This has clear consequences regionally and for the United States at home. In recognition thereof, the United States has used the Caribbean Basin Security Initiative to foster crime prevention programs in an effort to specifically address what the *International Narcotics Control Strategy Report* (*INCSR*) called "citizen safety." In one such initiative approximately 2,500 Caribbean police officers were trained in the Dominican Republic. These training exercises help build the capacity of law enforcement to meet the needs not only of investigating complicated crimes (for example, forensic accounting for international financial crimes) but also of mitigating the local fallout from transnational crime. The Department of State credits such mentorship programs in St. Kitts and Nevis, for example, with reducing homicides by 41 percent.[22]

These programs came in response to the drastically changing nature of Caribbean crime. The threat of local crime became so profound—indeed, so hybridized (the blending of crime and war)—that Caribbean law enforcement underwent its own parallel hybridization in the 1990s. In Jamaica, for example, Anthony Maingot tells how the fear of the economic impacts of violent crime prompted a heightened militarization of police command when an army colonel was seconded to the police.[23] Yet, while militaries frequently provide superior logistics and resources, this militarization of local police work was doomed to failure. As police departments experienced in the United States in the 1980s and 1990s, a combative posture frequently alienates the very communities that authorities need to reach. American law enforcement is yet again experiencing the fallout from increased militarization, this time in the wake of 9/11.

David Kilcullen arrives at a similar conclusion in his discussions about American military action in Mogadishu and Jamaican combined military and police action in the Tivoli Gardens garrison district.[24] Both are examples of urbicide, similar to my discussion of Israeli and Brazilian tactics earlier, in which decisions that may be tactically effective instead produce an environment that is toxic and ultimately strategically damaging. Considering the shortcomings of military tactics used in police contexts, community policing has attracted greater attention from Caribbean law enforcement. New York Police Department officers, for example, have deployed to the Caribbean, as in Haiti, to serve as advisors to local law enforcement.[25] Even the statistical software introduced by Commissioner William Bratton in New York City to map crime on a neighborhood level has become popular across the Basin. Both the Costa Rican Ministry of Public Security and the Panamanian National Police, for example, are implementing the COMPSTAT (comparative statistics) system.[26]

The adoption of technologies closely related to the broken windows theory is accompanied by the adoption of many community policing precepts in the Caribbean. The Association of Caribbean Commissioners of Police (ACCP) acknowledged community policing (of which broken windows is a subset) as its preferred strategy back in 2001. The ACCP has partnered with the Centre for Criminology and Criminal Justice at the University of the West Indies to help better understand and pursue community policing.[27] The Regional Police Training Centre in Barbados incorporated community policing in its courses, noting on a syllabus for recruit training that students would be "exposed to the concept of Community Policing."[28] More senior officers would also be eligible to enroll in a two-day seminar entirely devoted to the subject. In 2012 the Bahamian government launched Urban Renewal 2.0, an array of community policing programs aimed at tackling the schism between neighborhoods and law enforcement.[29] The United States likewise supports community policing initiatives in the Bahamas as well as in Belize, Costa Rica, the Dominican Republic, El Salvador, and Guatemala.[30] The Panamanian National Border Service is particularly noteworthy for its incorporation of community-oriented missions in addition to its traditional responsibilities, according to the State Department. On the conceptual level, community policing has certainly won the day.

Yet, despite the strong preference for community policing techniques among Caribbean law enforcement, a number of factors have slowed the actual implementation of community policing, particularly in an integrated manner. First, despite the organization's consensus, ACCP met difficulty convincing the Caribbean Community (CARICOM) to pursue an integrated approach to policing, according to North Carolina State University professor Clifford Griffin.[31] For reasons of geography and politics, he notes, countries have traditionally pursued individualistic and isolated enforcement strategies, independent of one another. ACCP thus joins the many other organizations Griffin chronicles—those providing assistance in criminal investigations training, anti–money laundering, narcotics enforcement, customs enforcement, and assorted multilateral groups like the Regional Security System and the Organisation of Eastern Caribbean States—as one of many organizations struggling to build unity of effort.

Second, local disillusionment with crime can at times perpetuate demands for heavy-handed rhetoric and action, as noted by University of the West Indies (Trinidad and Tobago) professor Ramesh Deosaran. This backlash is understandable and is explained in part by the small geography and demography of Caribbean states, which generate a distinct sociology in which even small spikes in crime can create "panic-driven concerns."[32] In this atmosphere, both "public mood and political expediency" can make the application of community policing difficult, according to Deosaran. Recall, however, that similar attitudes seen in chapter 2 were ultimately overcome when police successfully implemented broken windows policing. Moreover, perspectives in the Caribbean community on inadequate law enforcement and insecurity are the very malefactors that, according to the broken windows theory, contribute to high crime rates.

Crime in the Caribbean has become such a complicated and integrated part of the regional discourse on security that some scholars, Deosaran relays, have called for a distinct "Caribbean Criminology." However, I would suggest that there is no need for an entirely new paradigm on crime in the Caribbean. What is instead lacking is simply a widespread application of an existing criminological theory, like broken windows, on a regional scale. This has been advocated for, on a conceptual level, by ACCP, which unequivocally supports the community policing paradigm. An American maritime security strategy based on the broken windows theory can lend particular benefit here. As I noted earlier, Caribbean regional initiatives often require an external patron, one that necessarily shapes an initiative's objectives. A United States–led broken windows maritime security strategy offers a theoretical avenue for ensuring unity of effort across stakeholders, using a framework already popular with Caribbean practitioners.

An integrated approach to crime in the Caribbean is not simply about achieving economies of scale in enforcement. As Griffith writes, crime is undoubtedly a "regional security challenge," not simply a "challenge for states within the region."[33] Crime is simultaneously local and transnational in the Caribbean, requiring an approach that is regional in scale but local in outlook. From the above discussion, we can see that community policing offers a good foundation upon which to build broader consensus in the region, particularly if orchestrated and partially funded by the United States. There are also incentives for the United States to leverage community policing's popularity as an organizing principle of maritime security. As General Kelly noted to Congress in his 2013 posture statement, regional security concerns, left unaddressed, have the potential to rapidly escalate into threats to the homeland.[34] Local crime, a regional priority, is not divorced from transnational narcotics trafficking, an American priority. With the right theoretical approach, combating one means combating the other.

People

The crime goes by many names: trafficking in persons, involuntary servitude, slavery, forced labor, debt bondage, sex trafficking. While definitions vary across these terms, they are all characterized by the "act of recruiting, harboring, transporting, providing, or obtaining a person for compelled labor or commercial sex act through the use of force, fraud, or coercion."[35] Such a definition (that of the U.S. State Department, and similar in language to that of the United Nations) includes some acts that the word "trafficking" might not immediately conjure. For example, an individual need not necessarily be physically moved in order to be a victim of trafficking. Children under the age of eighteen, for instance, cannot legally consent to prostitution under U.S. (and much international) law. Thus, any child engaging in prostitution while employed in the services of a pimp or brothel is by definition a trafficking victim. Even those who move willingly may be victims of trafficking. If an individual is enticed to move based on fraudulent employment claims and is then held with confiscated passports or charged exorbitant fees for their relocation, that person is a trafficking victim regardless of whether they moved under their own volition. Human trafficking is also frequently confused with human smuggling. And there are similarities to be sure. Both frequently involve the movement of people in deplorable conditions, and both populations are vulnerable to abuse. Yet the primary difference between the two is that smuggled individuals acquiesce (for lack of a better word) to their movement, paying or compensating smugglers for their services. Trafficking victims, on the other hand, are universally characterized by their lack of consent (or inability to legally consent, in the case of minors). In this section, I explore the impacts of both of these vast, irregular movements of people throughout the Caribbean Basin and their interplay with the crimes and insecurities addressed thus far.

Human Trafficking

If some crimes could be said to be more vile than others, trafficking in people should be near the top of the list—not just because treating people like commodities is abhorrent but because those victimized are often among society's most vulnerable. To wit, in Central America and the Caribbean, children accounted for 58 percent of all documented trafficking victims for the most recent year on record (2016) according to the UN Office on Drugs and Crime's *Global Trafficking in Persons Report*.[36] And while the UN's 2012 report noted that trafficking in children was most common along the periphery of the Caribbean Basin, predominantly in Central America and the northern tier of South America, its 2016 iteration warns that trafficking in children is now an emerging trend for the broader Central American and Caribbean region.[37] With respect to gender, a wide majority of trafficking victims in Central America and the Caribbean are female (both women and girls). Fifteen percent of victims in Central America and the Caribbean, children included, were trafficked for the purposes of labor exploitation. When Latin America is included, the percentage of victims trafficked for the purpose of forced labor is, according to one Congressional estimate, larger than in both Europe and Central Asia.[38] In this broader region, profits from coerced domestic servitude are estimated at half a billion dollars annually and forced labor at one billion dollars.[39] The International Labour Organization (ILO) has estimated that 1.2 million forced labor victims may be enslaved in the region.[40]

The majority of detected victims in the Americas, however, are trafficked for sexual exploitation. Sex trafficking in Latin America is estimated to generate more than $10 billion annually, according to the ILO. The International Organization for Migration (IOM) has placed that figure closer to $16 billion annually.[41] The ILO estimates there are 400,000 victims of forced sexual exploitation in the region, which would average out to an annual profit of $27,000 exploited off of each individual sex-trafficking victim. Tragically, people evidently make good business.

Yet human trafficking has historically been regarded as less of an enforcement concern in the region compared to narcotics. Consequently, the penalties for human trafficking are frequently less than those for narcotics trafficking. Skyrocketing profits and lower risks (compared to narcotics) have driven the trade in human beings to new heights. Trafficking in persons is likely the fastest growing illicit trade in the world, by Naím's estimate, and the second-most prevalent transnational crime in the Caribbean after narcotics smuggling.[42]

Generally, trafficking in the Western Hemisphere is intraregional. Victims detected in Central America and the Caribbean overwhelmingly originated from Central America and the Caribbean. Just as with narcotics, however, the Caribbean Basin is also a channel for transshipment, including the interregional flow of Asian trafficking. Victims from East Asia make up more than 16 percent of total detected victims in North America according to the United States in 2016, many likely passing through the Caribbean.[43] In previous years, South and East Asian victims had accounted for more than a quarter of detected victims in North America, Central America, and the Caribbean (see the 2012 UN report, for example), and 10 percent in South America. Chinese victims have been found in Venezuela, Colombia, Mexico, and Ecuador, and East Asians in Costa Rica, El Salvador, and some Caribbean island states. On the other hand, Central American and Caribbean victims have been detected across Western and Central Europe and, to a lesser extent, in the Middle East and East Asia,[44] while the Middle East and North Africa may have overtaken Europe in recent years.[45] The 2016 *Global Report on Trafficking in Persons* records that victims from Central America and the Caribbean have now been uncovered in more than thirty countries around the world.[46]

As with legal migration, in both intra- and interregional cases, the general trend is "south to north" trafficking, from poorer to wealthier regions. In line with this, a large degree of victims detected in the United States originate from developing Caribbean Basin countries. Yet, south–north migration is common even on less extreme scales. Most detected trafficking victims in Mexico are Guatemalan, most trafficking victims in Guatemala are from El Salvador and Nicaragua, and many trafficking victims in El Salvador originate from Honduras (per the United Nations).[47] The Bahamas, which has a thriving tourism industry, demands low-skilled and low-priced labor, which is frequently supplied by irregular (sometimes forced) migration from Haiti, Jamaica, and Cuba.[48] Thus, while the Caribbean Basin is often spoken of as a source region in the trade of people, the intraregional flow is equally dynamic.

All of the above statistics are just estimates. As the Congressional Research Service's Latin American specialist Clare Seelke notes, "despite the relatively large number of victims trafficked for forced labor in the region, there are relatively few studies on the topic."[49] As a result, as with most elements of illicit trade, we are forced to extrapolate from the available points of reference. But figures on human trafficking are also ambiguous because of the sheer scale of individuals at risk. Seelke writes that the State Department suspects that at least 100,000 Latin Americans are trafficked across national borders annually, many from Caribbean Basin states such as the Dominican Republic and Colombia. The State Department has estimated that perhaps as many as one million children are employed as domestic workers in Latin America, leaving them highly vulnerable to coercion and trafficking.[50] The ILO has estimated that more than five million children work in Latin America and the Caribbean, many (perhaps most) in unsafe conditions.[51] While child labor may not always qualify as human trafficking, it serves as a significant risk factor for forced labor. This also operates on a family level. Children of parents in forced labor are significantly more vulnerable to forced labor and trafficking themselves, according to the ILO. The ILO explored this hypothesis in Guatemala, where more than 40 percent of those with jobs were recruited under deception or coercion, and where 18 percent of working children had parents who were victims of (or vulnerable to) forced labor. The ILO has also estimated that almost two million people across Latin America are victims of forced labor, many of whom may be (or may become) trafficking victims.

On an individual basis, the list of risk factors is extensive: "poverty, unemployment, illiteracy, a history of physical or sexual abuse, homelessness, drug use, and gang membership" all increase the likelihood of being trafficked, according to Seelke.[52] Even climate change and natural disasters are risk factors. More extreme weather occurring at greater frequency destabilizes communities and causes considerable dislocation. After the earthquake that struck Haiti in 2010, hundreds of thousands of orphaned children found themselves at risk of trafficking.[53] From a demographic perspective, children and women are most at risk. Women and children from the Dominican Republic and Guyana have been detected trafficked into Venezuela, while Venezuelan children have been detected trafficked into Guyana for sexual exploitation in the rural mining communities.[54] Seelke fingers Panama in particular as a destination for Colombian and Central American women trafficked for sexual exploitation. And while children are most frequently trafficked for sexual exploitation and domestic servitude, she underscores that they have also been victimized for "illegal adoptions, for use as child soldiers . . . and to work for organized crime groups."[55] Venezuelan children, for example, have been trafficked into Colombia and pressed into service as child soldiers for the guerillas operating there.[56] Regarding industries, the U.S. Department of Labor has reported that children in Latin America were coercively employed in the production of "bricks, gold, coffee, sugarcane, and other agro-export crops," while initiatives to counter forced labor and human trafficking in supply chains have gained increasing momentum.[57]

Beyond the statistical risk factors endangering potential victims, there is also the culture of movement in the Caribbean. On a per capita basis, the Caribbean Basin exhibits "extensive migration, with a rate of movement in relation to the population that may be one of the highest in the world."[58] Of the thousands of economic migrants transiting from Haiti to the Dominican Republic alone, many lack adequate documentation and are forced to resort to human smuggling. And while smuggling is not the same as trafficking, being smuggled is an associated risk factor for further victimization. The same criminal networks occasionally run both smuggling and trafficking operations, according to Naím, with the former easily morphing into the latter.[59] As Lara Talsma writes for the UN High Commissioner for Refugees, this places migrants at significant risk of trafficking, either through exorbitant smuggling fees or forced labor.[60] Migrants are at risk of kidnapping for ransom too. Kidnappings have the potential to escalate into trafficking if perpetrators sense an opportunity for greater profit. Kidnapped migrants have been forced to pay off their smuggling "debt" through forced labor, domestic slavery, sexual exploitation, or even recruiting other migrants for trafficking.

Ease of movement, which facilitates the tourism trade in the Caribbean, has particularly insidious implications for sex trafficking, which remains an endemic challenge. As Seelke notes, the IOM reported more than a decade ago that the region's open borders, poorly regulated prostitution industry, and generally liberal enforcement of entertainment visas and work permits make trafficking considerably easier.[61] Curacao, for example, allows prostitution by permitted nonnationals in the zone of Campo Alegre, but regulation is so minimal according to the IOM that illicit prostitution (and thus trafficking) is present outside the zone. St. Lucia provides yet another example in which loosely regulated entertainment visas may be related to illegal prostitution.[62] These lax travel restrictions have contributed to the ongoing rise of sex tourism in the Caribbean. The Dominican Republic, Mexico, Honduras, Costa Rica, and Trinidad and Tobago have all been identified as destinations of concern in this industry.[63] According to IOM interviews with individuals in Jamaica, Barbados, St. Lucia, the Netherland Antilles, and the Bahamas, all believed there was a connection between sex tourism and human trafficking in the region. The consequences of this trafficking are not localized to the individual victims either. In chapter 2 I note the relationship between the broken windows theory and public health research, including that on sexually transmitted infections. Human trafficking presents acute risks to public health through similar mechanisms. Victims forced into sex work are at significantly higher risk of contracting and transmitting HIV/AIDS and other sexually transmitted infections while disorder propels a wider culture of high-risk sexual behavior.

Yet perhaps the most relevant risk factor for trafficking, as relayed by Seelke, is the presence of preexisting transnational networks. These groups and the individuals composing them are at heart capitalists, and any opportunity to diversify "products" means both a greater chance for profit and more distributed risk. Like drug smuggling, human trafficking networks are frequently disaggregated, characterized by small groups temporarily collaborating in ad hoc arrangements. Consequently, the architecture for human trafficking is often shared with that of narcotics, just as narcotics trafficking and weapons smuggling are interrelated. Seelke assesses that criminal groups in Guatemala have expanded into human trafficking as a means of financing other illicit endeavors.[64] In Mexico, meanwhile, Los Zetas (a breakaway group of the Gulf Cartel) have become prominent in human trafficking. Mexican organizations have even partnered with other ethnic cartels, including Ukrainians and Chinese, to further diversify into human trafficking. "By chaperoning other shipments against a fee," Naím writes, narcotics networks have expanded into moving virtually every type of U.S.-bound illicit product. In fact, many organizations (particularly Mexican ones) originated as human smuggling networks before taking to narcotics.[65]

Victims are subsequently pressed into service for gangs and cartels to steal, produce, and smuggle drugs and even to kill, further blurring the lines between narcotics trafficking, human trafficking, and local violent crime. Reports in the *Trafficking in Persons Report* chronicle that Mexican organizations have forced children to serve not only as lookouts and mules but also as assassins. Informants to an IOM study in Jamaica noted that boys were used as lookouts for drug dealers, who also sexually abused them, while girls were sold into the local sex trade. The multidimensional dynamic of maritime insecurity in the Caribbean cannot be on greater display than in the relationship among crime, narcotics, and human trafficking.

Context is also at play, if more subtly. The impact of trafficking on communities, which the broken windows theory identifies as of compounding significance, is particularly devastating. A report from the IOM captures both context and multidimensionality, noting that victims signal the existence not only of trafficking rings but of the other criminal activities in which such networks are typically involved—from narcotics to weapons smuggling. Moreover, the predominance of these activities portends corruption in the ranks of government personnel as well.[66] Victims of trafficking therefore send a powerful signal to communities of disorder, violence, mistrust, and collusion. They signify a broken system.

Irregular Migration

Irregular migration (migration through unofficial, often illegal channels) is driven by an array of pull and push factors. Obvious pull factors include the prospect for employment or reuniting with family members already living abroad. A more subtle pull factor may also include favorable notions of migration, which the IOM assesses is "perceived very positively by Caribbean peoples" as a mechanism for progress. Push factors include much of what we have and will discuss in this section (crime, corruption, unemployment). In Central America, excessively high crime rates have pushed migrants northward to Mexico and the United States in growing numbers. Mexico deports more than 100,000 Central American migrants every year, many in a bid to reach the United States. Facilitating this surge are many of the same networks seen throughout this chapter, which have adapted considerably over the decades.

The term "coyote" remains popular as a label for those who help illegal migrants over the Mexico-U.S. border (the busiest in the world). Yet, in recent years greater enforcement efforts on the border and rising demand in Central America and the Caribbean have acted in concert to raise the cost of transit from around $300 in the mid-1990s to several thousand dollars in the early 2000s. Larger integrated networks have an advantage in coordinating more expensive and complex operations and, thus, in the words of Naím, the "old-fashioned coyote may soon be obsolete."[67] If the coyote were the "small-scale entrepreneurs," their replacements, *polleros*, are "professional smugglers."[68]

The longer the distance an individual must travel, the more likely they will need to employ the services of a criminal network, or *polleros*.[69] And people are moving from all over the world. Migrants from China's Fujian Province or Thailand, for example, rely on sophisticated networks for help in traveling to Central America or Mexico before continuing to the United States or Latin America.[70] Regionally, meanwhile, demand is spiking. Pulled by economic potential and pushed by criminal insurgencies, General Kelly testifies that migration from Honduras, Guatemala, and El Salvador rose 60 percent in 2013.[71] In fiscal year 2016, meanwhile, the United States recorded a further 23 percent rise in illegal border crossings, with

Central Americans highly represented.[72] Canada, the United States, and Mexico have all seen a rise in asylum claims from the Caribbean Basin, and most asylum claims in Mexico come from Central America, according to Talsma.[73] The United States witnessed a "44-fold increase in Haitian migrants in the Mona Passage" around the time Kelly was at the helm of Southern Command. Between 2005 and 2012, he notes, Southern Command detected a total of 200 Haitian migrants in the passage. In the first few months of 2013, on the other hand, more than 2,000 were detected.[74] Kelly's successor (Adm. Kurt Tidd) contended with a continued rise in the number of Haitian migrants detected on both land and sea, from 3,435 to 7,932 between 2015 and 2016.[75]

As with human trafficking, the lived context in which irregular migration takes place is important for understanding the phenomenon. Dire circumstances produce a pervasive desperation, which can push people to escape at any cost. Unemployment rates in the region are, according to Talsma, "among the highest in the world, up to 42.9 percent in Guatemala and 44.6 percent in Honduras." The concurrent spike in gang membership and criminal violence throughout the Basin has made things even worse. In the words of one Salvadoran migrant: "We are going to die, one way or another."[76] The sudden arrival of tens of thousands of Caribbean children on the U.S. southern border in 2014, sparked by nothing more than a circulating rumor, speaks to the panicked context in which this massive scale of migration takes place. Migrant smuggling networks expanded in the Caribbean at that time in part, Kelly told Congress, because "new laws in Cuba and erroneous perceptions in Haiti of changes in U.S. immigration policy" increased flows from the Eastern Caribbean.[77] Only months later, in early 2015, the announcement of an American–Cuban rapprochement triggered yet another massive wave of Cuban migrants, again aided by a rumor (this time the repeal of the "wet foot, dry foot policy," which was indeed revoked at the end of President Obama's term). As the broken windows theory contends, context (both real and perceived) plays an important role in decision making, with real consequences for authorities. The U.S. Coast Guard, for example, interdicted nearly 10,000 Cubans, Haitians, and Dominicans attempting to enter the United States by sea in fiscal year 2014 (the prior year, that number was around 7,000).[78] Migrants, many Haitian (often the largest demographic among those interdicted), are also increasingly transiting through the Dominican Republic to take small boats to Puerto Rico in a passage that is evocative of Hispaniola / Puerto Rico narcotics flows (with people and drugs occasionally even using the same boats).[79]

Subregional irregular migration in the Caribbean Basin takes place for much the same reasons that migrants flow to Mexico and the United States.

Uneven prospects for employment and varying levels of governance act to direct Caribbean migrants across the Basin, often in illicit ways. Across the region, gross domestic product per capita is as low as $1,000 in Haiti, and as high as $20,000 in Trinidad and Tobago. Life expectancy ranges between sixty-one and seventy-eight, while adult literacy can be as low as 62 percent and as high as 100 percent.[80] Barbados, for example, has strong literacy and per capita income numbers and ranks high in the UN *Human Development Report*'s Human Poverty Index, thus making it an attractive destination for irregular migration.[81] Bermuda, Puerto Rico, and Trinidad and Tobago, considered "borderline countries" (between developing and developed), have also been enticing destinations, as have been the Bahamas, Guyana, Jamaica, the Netherland Antilles, and St. Lucia.[82] It is within this flow of people that human trafficking is nestled deeply, and the undocumented status of irregular migrants places them at heightened risk of trafficking. The cons used to deceive trafficking victims—often employment offers—are made all the more believable by this pervasive irregular migration pattern.

The Bahamas, whose geographic position between Florida, Haiti, Jamaica, and Cuba makes it a hub for such activity, offers one such example. Small boat and plane captains are alleged to be in frequent transit between Hispaniola and the Bahamas, advertising construction and hospitality jobs in the latter.[83] The IOM relays how industrious Haitian boat owners charter migrants to the Bahamas for a cost, all to serve the needs of the growing tourism sector. While such trips may not be coerced, in the most direct sense, they are nevertheless very dangerous. Boats are frequently loaded beyond capacity, and reports of migrants drowning in sinking vessels are tragically common. After three successive fatal incidents involving individuals from Haiti, the Dominican Republic, Cuba, and Jamaica in 2013, the office of the UN High Commissioner for Refugees "expressed its official concern" over the plight of economic and political migrants. In one incident, more than one hundred Haitians were crammed into an "overloaded sail freighter," which ran aground off of the Bahamas' Staniel Cay. Thirty migrants died before the U.S. Coast Guard and Royal Bahamian Defense Force arrived. This tragedy was part of a wider trend in the northern Caribbean at the time, which saw a steep rise in maritime deaths between 2012 and 2013.[84]

Just as with human trafficking, human smuggling and irregular migration present several challenges for communities. I have already addressed how smuggling places migrants at risk of trafficking, which opens communities up to the exploitative schemes of transnational crime. I have also discussed how illicit networks are multidimensional, opportunistic, and capitalistic, expanding into a range of interconnected trades. General Kelly puts it concisely: "Clearly, criminal networks can move just about anything on [their] smuggling pipelines."[85] Criminal networks may even use human smuggling to swell their ranks. The deputy commissioner of police in Guyana once voiced concern that "foreign mercenaries" were augmenting gang membership in the country.[86] There are similarly public health risks that accompany the mass underground movement of people. In the midst of the 2014 Ebola epidemic in West Africa, the commander of Southern Command told an audience at the National Defense University that the spread of the disease in Central and South America was on his mind. Haiti and Central America were of particular concern because of poor health care and their place in the irregular migration chain. General Kelly noted especially that the individuals smuggled by transnational criminal networks into the United States may be infected and enter without medical oversight.[87] This fear may not be merely theoretical. The region has a history of epidemiological tragedy; the Caribbean is in fact the hardest hit region in the HIV/AIDS epidemic after sub-Saharan Africa.[88] In a visit to the Costa Rican–Nicaraguan border, General Kelly recounts meeting several African migrants. The men told U.S. embassy personnel they were Liberian and had been traveling for a week, with the aid of a network from Trinidad, to the United States. The general noted that those men could have made it all the way to New York City within Ebola's incubation period.[89] And while fear of Ebola in the West was almost certainly overblown, the premise is nonetheless significant. Smuggling helps transport not only people, narcotics, and goods but also diseases that can ravage the underprepared littoral and small island states of the Basin.

Finally, there can be no greater metaphor for the relevance of the broken windows theory's attention on environment than the impact of climate change on local communities. As climate change becomes an issue of greater concern for Caribbean islands and littoral states, issues of irregular migration and human trafficking will only become more significant. Storm surges, flooding from changing precipitation patterns, extreme weather, coastal erosion, and droughts will all, in the words of the Coast Guard's *Western Hemisphere Strategy*, "disrupt trade routes, threaten lives, and drain national economic resources," pushing climate refugees into greater motion. The demographic factors addressed in chapter 1, the *Strategy* continues, "coupled with climate change, will impose extreme stress in the region."[90] The 2010 earthquake in Haiti, while not related to climate change, illustrates just how much upheaval natural disasters can bring to bear on irregular migration patterns. Even efforts to combat drug smuggling, such as the crop-dusting of wide swaths of farmland, can damage the environment and instigate irregular migratory flows.[91] As Caribbean communities continue to respond to the stimuli of their environment, any theory of maritime security must take pains to acknowledge and address the context within which these lives are lived.

Finally, of those irregular migrants and asylum seekers detected during passage, many are repatriated to their countries of origin. While repatriation is not an illegal flow, this irregular migration pattern is nonetheless noteworthy. The Bahamas, for example, as a regional destination, frequently repatriates Caribbean nationals to their countries of origin, chief among them Haiti and Jamaica. Migrants from Hispaniola make up such a significant proportion of irregular migrants that Haiti and the Dominican Republic are the only two Caribbean countries for which nationals are required to procure a visa before entering at least one Caribbean nation (Barbados), as noted by the IOM.[92] Yet perhaps the greatest repatriation flow comes in the form of deportations. The high gang membership figures noted earlier are owed in large part to this practice. Some estimates, like that relayed by Talsma, hold that as many as 70 percent of gang members arrested in the United States are eventually deported, most to the Caribbean Basin. Consequently, gang networks have gone multinational—so much so that antigang investigations in the Caribbean have direct relevance for American law enforcement. The State Department's INCSR attests to this, with some cases originating or pursued in the Basin leading to "a homicide arrest in Oklahoma City, the prosecution of felony extortions in Annapolis and the capture of one of the FBI's top ten most wanted fugitives ... in El Salvador."[93] These gangs are tethered to affiliates across Latin America because of such widespread deportation. As Kelly told Congress in 2013, "MS-13 [Mara Salvatrucha] gang leaders in El Salvador have initiated assassination plans against U.S. law enforcement personnel" and targeted Americans throughout the region.[94] The Virginia gubernatorial election in 2017, replete with campaign advertisements targeting fears of MS-13, demonstrates the persistence of the perceived threat of multinational gangs in American society. The group has a particularly strong American presence in California, North Carolina, New York, and Northern Virginia, among other major cities. MS-13, in Kelly's estimation, "specializes in extortion and human trafficking" and has succeeded in part because deported members help provide international reach.[95] Honduras has suffered exceptionally from deportations. Migrants are sent back en masse from American prisons, many who are members of Barrio 18 and MS-13. And in testimony before Congress, General Kelly relayed the estimates of the National Drug Threat Assessment that Mexican-based transnational criminal organizations now operate in at least 1,200 American cities. Deportations are such a staple of the American criminal justice system that, despite sour relations on most issues between the United States and Venezuela, the two coordinate on two principal missions, according to the State Department: maritime narcotics interdiction and fugitive deportation.[96] Yet this is not solely an issue of American provenance. Intraregional criminal deportation is also a frequent occurrence. Barbados, which I noted was a borderline country and the target of regional irregular migration, allegedly deports around twenty Caribbean nationals a week, principally to Guyana, St. Lucia, St. Vincent and the Grenadines, and Trinidad and Tobago.[97] Deportation and repatriation thus serve not only to facilitate irregular migration but also as contributors to transnational and local crime. More broadly, the movement of people in the Caribbean is intertwined with local crime as well as the movement of narcotics and weapons. It is increasingly clear that we cannot address one without addressing the others. It is also increasingly clear that we cannot address any of these issues without considering the lived environment of those most affected.

Obstacles
Responding to the implications of mass human movement in the Caribbean Basin is a monumental challenge. As human trafficking continues to grow faster than any other illicit trade, as human smuggling continues to transport people and diseases across borders, and as the networks that perpetrate these deeds expand in scope and "product," irregular migration in the Caribbean is an issue of security concern. Yet the obstacles to any strategy aimed at this task are manifold. First there are the economic disincentives to greater enforcement. As the Caribbean attempts to further regional integration, barriers to human trafficking and irregular migration will continuously come in conflict with efforts like CARICOM's Single Market and Economy initiative. While the European Union's Schengen Zone–inspired freedom of movement is far from a reality in the Caribbean, many in the region aspire to such a community "without barriers."[98] Economic incentives favor the open exchange of goods and people, and U.S. direct investment in Latin America at large exceeds $1 trillion (second only to U.S. investment in Europe), according to the Coast Guard. Of the twenty states with which the U.S. government has signed free-trade agreements, "more than half are in the Western Hemisphere."[99] Furthermore, the industry that drives the demand for so much cheap labor across the Basin, tourism, is in its own right the source of the largest movement of people in the Caribbean (more than twenty million visitors annually).[100] On top of tourism, the constant flux of domestic migration flows add to an ever-dynamic movement of people. In the IOM's judgment, this extensive domestic migration makes it

all the more difficult to detect victims of trafficking as well as irregular migrants. The most obvious obstacle, however, is access to resources.

If counternarcotics operators believe they receive a fraction of the resources they require, antihuman trafficking and human smuggling operations receive even less. According to Seelke at the Congressional Research Service, American funded, anti-human-trafficking-specific initiatives in Latin America totaled less than $8.5 million in fiscal year 2011 and about $11 million in 2016.[101] More than 60 percent of the funds in 2011 went to one country (Haiti), and about half went to a single country in 2016 (Guatemala). The increases in migration flows exhibited in the first half of this decade will, as Kelly told Congress, "place additional burdens on already overstretched U.S. Coast Guard" and Caribbean assets.[102] What action has been taken is done in the same vein as narcotics interdiction. But such enforcement must come as part of a broader strategy that recognizes that trafficking is not an isolated act. I highlighted earlier the interplay between human and narcotics trafficking. The multidimensional nature of human trafficking often means that victims are also implicated in local crime (prostitution or theft, for example); victims are often misidentified as criminals or illegal migrants.

Engagement on issues of human rights and human dignity is a sensible cornerstone for a strategy guided by broken windows. As the theory illustrates, a population's psychological well-being is instrumental in building communal security and interrupting individual criminal actions. Southern Command is acutely aware of this need as well as of the difficulty in building and sustaining goodwill, noting that the command's "persistent human rights engagement also helps encourage defense cooperation, trust, and confidence, which cannot be surged when a crisis hits, and cannot be achieved through episodic deployments or chance contacts."[103] Yet budget cuts have targeted the very missions necessary to maintain human security and human rights. Sequestration resulted in the cancellation of USNS *Comfort*'s deployment, for example. And goodwill is a competitive commodity in the Caribbean. Failure to deploy hospital ships to needy communities comes in opposition to an increasingly involved China and contributes to the disintegration of public health. Local competition is also stiff. Cuba sends nearly 30,000 medical professionals (mostly to Venezuela) every year; the United States, by contrast, sends approximately 700 to the entire Basin.[104] Thus, instead of offering the "next generation of global citizens direct experience with the positive impact of American values and ideals," in Kelly's words, Southern Command's resources for humanitarian assistance are shrinking.[105] And while broken windows theory reminds us that it is important that someone cares for a community, the proliferation of nonstate actors seen in chapter 1 reminds us that who cares for a community is just as important.[106]

Terror

No conversation on maritime security would be complete today without mention of terrorism. As I noted in the introduction to this section, terrorism is frequently seen as an issue at the periphery of the Caribbean, with Colombia the exception. Yet, because of the multidimensional character of maritime insecurity explored throughout this chapter, terrorism remains an issue of concern for Caribbean security. In fact, in many ways an assessment of terrorism in the Basin is emblematic of how the individual sections discussed above are even more tightly related.

The robust infrastructure for moving people and goods in the Caribbean is at potential risk of co-optation by terrorist organizations. Kelly warned Congress that the same networks that facilitate human smuggling and narcotics trafficking could as easily be leveraged by terrorists. This sentiment was echoed by Rear Adm. Charles Michel in his testimony to Congress in 2012 and by Admiral Tidd in 2017.[107] By forging relationships with transnational criminal networks and establishing their own, terror organizations could absorb the capacity to exploit undergoverned spaces, just as do regional transnational criminal organizations. Terror operations launched against the United States or Europe could be concealed and facilitated in the Caribbean for the same reasons illicit trafficking passes through the Basin: prime geographic and political positioning. The Caribbean is a "strategic gateway" to North America, South America, and Europe, considering the variety and volume of sea and air travel crisscrossing the region.[108] The political relationships maintained by the United Kingdom, the Netherlands, and France provide avenues for migration and transportation that further facilitate the movement of people and products. The CARICOM Regional Task Force on Crime and Security notes that terrorists could use the region's lax banking regulations, air- and seaport facilities, and vast shipping networks to facilitate attacks or terror financing.[109]

Of course, instances of direct terrorist attacks are scarce. There are also many downsides to discourage criminal networks working with terrorists—the calculated gain would likely pale in comparison to the increased scrutiny that criminal groups would attract in cooperation with terrorists. Notwithstanding these points, limited evidence, informed speculation, and less-informed accusations of international terrorist organizations operating in the region abound. Drug traffickers in the Caribbean Basin have been documented selling transport to individuals from outside of Latin

America seeking to enter the United States illegally, for example, which presents possible concerns of terrorists purchasing passage on smuggling pipelines even without the direct knowledge of the criminal organizations that operate them.[110] Hezbollah in particular has been accused of engaging in human trafficking in an effort to fund its operations, which demonstrates the technical capacity for human smuggling as well. A recent U.S. court case, *United States of America v. Anthony Joseph Tracy*, highlights both the ease with which such pathways can be used by terrorists and the difficulty of gathering enough intelligence to fully substantiate such claims.[111] Anthony Joseph Tracy, an American-born man, converted to Islam while in prison. As the leader of an international human smuggling ring based out of Kenya, Tracy admitted to facilitating the emigration of nearly three hundred Somalis to the United States over seven months in 2009. Using fake documents, Tracy would obtain visas for his clients to travel to Cuba, booking round-trip flights to demonstrate an intent to return to Kenya. Two such clients, KA and DH (Somali men in their early twenties), traveled to Cuba from Kenya via Dubai and Moscow. From Cuba, their trail becomes muddled, transiting South and Central America to arrive in Mexico, where they successfully crossed the border into the United States. Tracy facilitated his business through a contact in an unnamed South American embassy, where he secured at least three visas for Somali men. While Tracy maintained that he did not knowingly facilitate the transport of terrorists into the United States, he does admit that he was approached by al-Shabaab for the use of his services. So, while no available evidence suggests any widespread collaboration between terrorist organizations and human smugglers in the Caribbean, analysts like Seelke note that they have not ruled out the feasibility of such a scenario.

Anti-terrorism efforts are likewise complicated by the Caribbean migration context I have already described. The need to generate revenue in such resource-poor countries means that Caribbean states are incentivized to make freedom of movement a priority, which has facilitated trafficking in people. This culture of movement is no more evident than in the practice of selling citizenship. Merely by demonstrating sufficient assets, some Caribbean islands will grant long-term residency visas (which attracts criminals, and perhaps terrorists).

There are legitimate uses of this service, to be sure. The freedom of movement this offers can be a lifeline for individuals whose countries or nationalities impose upon them severe travel burdens. Hadi Mezawi, a Palestinian man living in the United Arab Emirates, obtained full citizenship in the small island of Dominica (without ever having been there) for the sum of around $100,000. Such "investor visas" are not exclusive to the

Caribbean, but the very low barrier for entry makes the region a popular choice, as explored by David McFadden.[112] Nowhere is the process faster and more worrisome than in Dominica and St. Kitts and Nevis. A passport from Dominica or St. Kitts and Nevis does not require an advance visa in more than one hundred countries. And St. Kitts and Nevis can process an application from start to finish in three months, without the applicant ever visiting the island. Such programs are so popular that a firm in Dubai was developing a community in St. Kitts and Nevis where individuals can "buy citizenship and property at the same time," according to McFadden's reporting. While, I underscore, there is no evidence of such programs being used by terrorists, the vulnerability of these visa and citizenship offers to nefarious use is one that deeply concerns governments. Canada, McFadden writes, has long placed visa requirements on Dominica passports over concerns that transnational criminals employed them. In 2010 Britain considered doing the same. Even Caribbean officials recognize the risk. Despite supporting the program, a St. Kitts and Nevis opposition leader, Mark Brantley, conceded that the government does not have "sufficient controls" in place to ensure that "bad people, for want of better language, do not get access to citizenship."[113] Hinting to these same fears, Grenada suspended its program after September 11, and St. Kitts and Nevis closed its program to Iranians (previously a major applicant pool) in the wake of an attack on the British Embassy in Tehran in 2011. Nevertheless, McFadden reported in 2013 that Antigua and Barbuda was launching a new citizenship program, and Grenada was considering reviving its own (which it eventually did).

Despite the theoretical potential for terrorist attacks staged in the Caribbean, many experts agree that there remains little conclusive evidence that any such plans exist. That is not to say, however, that there is no terrorist activity in the Caribbean at all. While evidence of plans for terror attacks are scant, the same cannot be said for evidence of terrorist financing and support. The Lebanese group Hezbollah, listed as a terrorist organization by the United States (and its military wing labeled a terrorist organization by the European Union), offers the most sophisticated example. Hezbollah established a presence in the tri-border area of Paraguay, Brazil, and Argentina in the mid-1980s and has since grown into a significant regional player.[114] Soon after 9/11, analysts hypothesized that sympathizers in that region had raised upward of $50 million in support of known or suspected terrorist organizations (dominantly Hezbollah). John Cope and Janie Hulse, writing in Griffith's anthology, also note that the region hosts "support activities" for a variety of Islamist groups, not all of which are terrorist organizations.[115] And while the tri-border region specifically is beyond the scope of this case study, evidence suggests that

Hezbollah has been financed in part by narcotics trafficking throughout Latin America. In May 2013 this was reinforced when Assad Ahmad Barakat, the head of Hezbollah's activities in the tri-border area, was arrested in possession of cocaine, which he intended to sell in the Middle East to help fund the organization. His arrest verified that a "wing of narco traffickers," in the words of Cope and Hulse, was in fact operating in Latin America in support of Middle Eastern terror.[116] The former U.S. ambassador to the Organization of American States, Roger Francisco Noriega, noted that Hezbollah was involved in a diverse array of criminal enterprises in Latin America, including narcotics smuggling, gunrunning, human trafficking, money laundering, and counterfeiting. According to figures from the Naval War College reported by CNN, these endeavors generate approximately $10 million annually.[117] Worldwide, in a considerable shift from the 2002 foreign terrorist report, the State Department now asserts that at least twelve of the world's twenty-five largest terrorist organizations have ties to narcotics trafficking.[118]

Iran's role, designated by the United States as a state sponsor of terror, has also spurred speculation in the region. An Argentine special prosecutor investigating the 1994 bombing of a Jewish community center in Buenos Aires has accused Iran of establishing terrorist networks "throughout Latin America."[119] The report in which this allegation was made names the Caribbean Basin states of Suriname, Trinidad and Tobago, and Colombia as possible locations with an Iranian presence. Alleged Iranian collusion with narcotics trafficking and access to the Caribbean through partners like Venezuela only serve to fuse terror, transnational crime, and international relations in still more complex ways. Venezuela surfaces routinely as a potential nexus where the worlds of narcotics, terrorism, and politics collide. As Kelly himself noted, two Venezuelans were sanctioned in 2008 in relation to Hezbollah financing.[120] And while Hezbollah may be interested in securing financial support in Latin America, some indications suggest Iranian factions see the region in more violent hues. News reports of an Iranian plot to assassinate the Saudi ambassador in Washington, D.C., in 2011, for example, involved a suspect soliciting aid from an agent posing as a Mexican cartel member.

Not only is it important to consider any relationship between terrorism and narcotics trafficking, it is even more significant to note that the two are not necessarily mutually exclusive. Some politically motivated organizations and individuals, as Cope and Hulse argue, "out of necessity or convenience" engage with narcotics traffickers as a means to fundraise, while still others "circumvent the intermediary and deal directly in the drug business."[121] Thus, while transnational crime is recognized as a far greater threat to the Caribbean Basin than terror, the nexus between the two is not to be ignored. The Revolutionary Armed Forces of Colombia (FARC) and the United Self-Defense Forces of Colombia (AUC) offer prime examples of how the line is blurred between terrorist and trafficker. FARC, staged in Colombia's hinterland, began as an ideologically motivated left-wing terror organization. The group's original intersection with narcotics was seen as a mechanism to finance a political cause, which was accomplished by taxing or selling protection to local coca growers. As the organization began to operate greater effective control over territory, its capacity to exact greater revenue from the trade enhanced, and its mission slowly shifted from political to capital. Simultaneously, FARC came under increasing pressure in the 1990s as right-wing paramilitary forces organized in opposition, raising the cost of operations.[122] By the middle of the decade, FARC had largely solidified its position as the intermediary between coca farms and cartel-run processing labs.[123] FARC soon leveraged its extensive territorial control to directly coordinate coca growth and even its own cocaine production, the products of which contributed as much as 50 percent to FARC's budget, Naím reports. The Colombian government estimated FARC earned nearly $800 million in cocaine revenue annually in the years after 9/11.[124] One news report placed that figure closer to $700 million, while Rhea Siers (of George Washington University's Elliott School of International Affairs) relays more conservative estimates between 75 and 100 million dollars annually.[125] The remainder of its budget was supplemented by kidnappings for ransom, sexual exploitation, and related extortive and criminal schemes. To borrow a phrase from Naím, with profits like these, "rebels turn into traders" and political objectives are left as veiled excuses for the pursuit of financial gain. In other words, drugs became the FARC's raison d'être.

In some respects, the AUC is a mirror image of FARC. The AUC was formed in 1997 as a right-wing response to FARC, when drug barons across Colombia banded together to create a network of militias to defend themselves against militants. The AUC was thus, from the outset, an instrument of narcotics trafficking, with its roots in the cartel armies of the 1980s.[126] The group reached a peace settlement with the Colombian government in 2003, but until that time, estimates suggest that 70 percent of the AUC budget was supplied by cocaine revenue.[127]

The Colombian government completed a peace process with FARC in 2016, resulting in a rapid drawdown of FARC forces after years of gradual force depletion for the rebel group. Reconciliation, however, may not portend exclusively good news for narcotics trafficking in the Caribbean. The particular skill sets associated with terrorism are often equally suited to a variety of criminal activities. When terrorist organizations wane, it is inevitable that at least some former terrorists, with years or decades of professional expertise in illicit operations, will shed their ideologies and gravitate toward crime for profit. This trend can be seen globally. The Provisional Irish Republican Army (IRA), for example, largely eschewed the world of drugs for the duration of its armed campaign against the British. The group frequently operated as vigilantes and did, in fact, help keep the burgeoning drug trade of the 1980s and early 1990s out of Belfast. Alongside cease-fire agreements in the mid-1990s, however, came a steady rise in Northern Irish drug consumption, aided both by the dissolution of the IRA's armed opposition to the trade and the prevalence of skilled and available labor. Remnants of the IRA itself turned to the smuggling of illicit goods and counterfeits to help sustain the group.[128] The IRA is also suspected to play a major role in the Dublin heroin market, and some relationship existed between the IRA and FARC, according to Naím's recounting.[129]

In some instances, terrorist organizations move en masse from terror to crime, saving foot soldiers the hassle of finding work on their own. For example, the Abu Nidal Organization, a Palestinian terror group, evolved from a largely political agenda to one increasingly defined by weapons trafficking and various petty criminal enterprises, "even a coupon fraud scheme," in Siers' estimate.[130] In Colombia, the National Liberation Army (ELN), a left-wing paramilitary group founded in the same year as FARC, evolved down a similar path. At its inception, the ELN was regarded as even more politically motivated than its ideological competitors (FARC). For that reason, like the Provisional IRA, the ELN chose to remain largely outside of the narcotics business. Since cresting in influence and membership in the 1990s, however, the group has increasingly allied itself with criminal gangs that principally serve the drug industry, according to reporting by the BBC. And while terrorist and paramilitary violence have racked Colombia for half a century, it is these criminal bands (*bandas criminales*, or *bacrims*) that are now regarded as the country's greatest rising threat. These groups have moved in to fill the void left by the stream of demobilized paramilitary groups in the country. In 2010 the Colombian think tank Indepaz reported that approximately a dozen new *bacrims* had emerged across Colombia in place of the AUC and were by then the cause of even more violence than FARC.[131] These criminal bands, despite inheriting some personnel of former terrorist and ideological movements, display no obvious political agenda. This path may illustrate a potential evolution for parts of Hezbollah as well. When financing from Iran waned in 2006, the organization became far more active in illicit trades in Latin America and West Africa in an effort to compensate. While the group still maintains its ideological agenda, Hezbollah is diversely involved with transnational crime in the Caribbean Basin, including formerly with FARC. In 2008, for example, Operation Titan (a joint U.S.–Colombian investigation) uncovered a cocaine ring directing profits to Hezbollah. In 2010 a Lebanese man was arrested selling cocaine to Los Zetas to fundraise for Hezbollah as well.[132] Mexican cartels are also believed to have partnered with Hezbollah in the trafficking of precursor chemicals for the manufacture of methamphetamine.[133]

While the relationship between terror and narcotics was once predominantly thought of as theoretical, the last few decades have shown the relationship to be genuine in the estimation of many analysts. At a DEA symposium held in Washington, D.C., only two months after 9/11, then-administrator Asa Hutchinson noted the "extraordinary link between drugs and terrorism."[134] To illustrate this point, DEA assistant administrator for Intelligence Steven Casteel noted that, two days prior to 9/11, law enforcement had "seized 53 kilos of Afghan heroin in New York. It was being distributed by Colombians, to show you the [potential] link."[135] Thus we come full circle. Terrorist organizations have the capacity to leverage the smuggling and trafficking architecture of the Caribbean to fundraise or stage attacks. Partnerships with criminal gangs link terrorism to local crime as organizations like Hezbollah or FARC operate at the intersection of crime and terror. Outgrowths of such organizations may even at times fuse with the narcotics industry, becoming more invested in profit than ideology, shifting the landscape of regional counterterrorism.

As terror organizations search for financial independence in the wane of state sponsorship, the financial relationships between these two worlds are increasingly significant. Congressional Research Service analyst Raphael Perl notes that both terror groups and transnational organized crime function best in an environment of intimidation and fear, creating a climate in which people feel helpless and abandoned by traditional security guarantors.[136] In this critical dimension, there is virtually no difference between transnational crime and terrorism. And in a theoretical construct like the broken windows theory, which takes aim directly at context, counterterrorism and counternarcotics (no matter how complex the underlying relationship) can and should pursue the same objective: changing the culture of Caribbean security. I explore that idea further in the final chapter of this section.

Lt. Cdr. Clifford Rutledge, a chaplain assigned to the commander of Naval Surface Squadron 14, takes a selfie with students at Republica de Colombia Elementary School during a Southern Partnership Station 17 (SPS-EPF 17) community relations project. SPS-EPF 17 is a U.S. Navy deployment executed by U.S. Naval Forces Southern Command and U.S. Fourth Fleet, focused on subject matter expert exchanges with partner nation militaries and security forces in Central and South America. *U.S. Navy photo by Mass Communication Specialist 1st Class Jeremy Starr/Released*

CHAPTER FIVE
Context and Conclusions

The last two chapters have demonstrated the interconnectedness of local crime, transnational trafficking (in people, drugs, and goods), and terror. If theory is useful to governments in helping to prioritize the allocation of scarce resources, the existing theories for security in the Caribbean Basin fall far short. In the narcotics context, I explored how Coast Guard, naval, and homeland security personnel are routinely underfunded for blanket patrols and interdictions—so much so that, according to both Gen. John Kelly and Rear Adm. Charles Michel, three-quarters of high-confidence trafficking incidents go unchecked because the ships and aircraft required to interdict them do not exist. Partnerships, which help distribute this massive cost, are slow in forming and easily dissolved, in part because American interests in Caribbean security do not always reflect the primary regional political and community concerns. Yet, as I explored in the section on crime, local incidents are inevitably part of a wider fabric of criminality and insecurity. Efforts to mitigate irregular migration, meanwhile, are even more underfunded than narcotics but as corrosive to communities and regional order. Terrorism touches on each of these worlds, especially so when security is seen in a multidimensional framework. As we can already see, signals of disorder abound in the Caribbean Basin's maritime environs, while a midcentury policing-style interdiction model dominates, despite having fallen out of favor in communities across the United States decades ago.

The lessons from broken windows theory on crime's multidimensionality, which we uncovered in chapter 2, hone our attention to the intersectional quality of Caribbean maritime disorder. There was, however, a second major lesson we gleaned from an investigation of the broken windows theory: the influence of environment. While criminal multidimensionality was largely observed as a consequence of broken windows' implementation, the theory's foundational core hypothesized that context was the most important factor in understanding criminal behavior. This supposition led to broken windows' prescribed enforcement technique: by combating signals of disorder, police could change how communities relate to their environment, building up communal self-efficacy and driving down crime. In other words, by targeting context (perceptions of crime, physical decay, appearances of governmental abandonment, pervasive acts of petty crime) authorities could influence crime rates more effectively than the midcentury policing model. We can see elements of this at play already in the preceding two chapters. I underscored, for example, the importance of context when understanding local perspectives on security. Context also plays an important role in understanding the consequences and efficacy of enforcement efforts (specifically, the discussion on urbicide, to which I return below). Put more colloquially, misunderstanding or ignoring context is precisely how militaries can win the battle but lose the war.

Yet context, as I explore in chapter 2, is far from narrowly defined. Both George Kelling and Malcolm Gladwell proffer definitions of context that do not restrict us simply to observable disorder but include perceived disorder as well. Recall, for example, the study on attitudes toward school safety, which coded disorder by how children *felt* about their schools, or the seminary students about whom Gladwell wrote. Such research underscored the psychological importance of how we internalize our environment and how that feeds into our behavior. Given this fundamental significance of the *perceived* environment to broken windows literature, I shift my focus in this remaining Caribbean case chapter to address two factors that heavily influence perceptions of the Caribbean security environment—money laundering and misgovernance—before turning to explore the broader lessons learned from our time exploring the Caribbean.

Money Laundering and Misgovernance

As we have seen throughout this part of the book, money is a core motivator for criminal enterprises and their drive to differentiate the "products" they offer. Yet making money is useless without a means to safely spend and enjoy ill-gained profits. Consequently, we cannot fully understand the impact of crime on local communities without exploring the mechanisms by which illegal revenue is laundered and, more broadly, the political systems that lubricate the process.

Money Laundering

In light of continuous resource shortfalls and the devastating threats posed by natural disasters, which can quickly erase a year's gross domestic product, small island nations have pursued a variety of creative opportunities for financial gain. Tuvalu, in the Pacific, is perhaps even infamous for this. Moisés Naím notes how the state makes a considerable portion of its revenue from licensing its domain tag, .tv, to major websites (such as MLB.tv).[1] The small island has even leased its country telephone prefix for use by a phone-sex company. Tuvalu also has a booming international finance industry despite being a virtual microstate. In fact, island states across the world, particularly in the Caribbean Basin, have become renowned financial centers. All that is necessary is a small shelter with enough electricity to power some servers, and an international bank is born.

Just as the easy movement of people facilitates greater revenue for the Caribbean's tourism sector, so too does the easy movement of money facilitate a lucrative banking sector. And just as the easy movement of people obscures and facilitates myriad trafficking and smuggling operations, so too does the easy movement of money. Lax banking restrictions breed corruption, while tax havens can attract criminal elements alongside your garden-variety tax dodger.[2] Of the major money-laundering countries identified in the 2016 *International Narcotics Control Strategy Report*, one-quarter are Caribbean Basin nations and territories. The Caribbean is recognized as such a focal point for money laundering that, in one year (2013), fully 80 percent of the UNODC's Global Programme against Money Laundering training participants were Caribbean states and territories.

Perhaps the most well-known offshore banking destination is the Cayman Islands. This country has one incorporated bank for nearly every one hundred people, not to mention mutual funds numbering in the thousands and incorporated offshore businesses in the tens of thousands, by Naím's count.[3] In the vortex of illegal money, Zackrison writes that the Cayman Islands' financial system is notorious for providing perhaps the best "climate" for laundering, facilitating the cleansing of as much as ten billion dollars in criminal proceeds annually.[4] Other states, competing for the windfall of cash from offshore financing, provide an equally liberal climate for laundering. St. Kitts and Nevis, for example, permits the establishment of corporate trusts designed to completely conceal the identities of proprietors.[5] And the use of the word "climate" here is noteworthy. Our language (as I explore more directly in later chapters) frequently evokes references to environment or culture when describing things like graft or corruption. This word choice is instructive, especially insofar as it may be an instinctive choice for authors and not deliberate, speaking to the fundamental yet unconscious associations we draw between a security environment (the word is inescapable) and the way it makes us feel.

The same characteristics that make the Caribbean attractive for narcotics and human trafficking also make it appealing for cash smuggling and money laundering. Zackrison chronicles a few of the more important factors, including a modern IT infrastructure, a large English-speaking population, expatriate Westerners living on the islands, émigrés living in the United States and Central America, geographic proximity to both North America and South America, access to Europe, and the presence of transnational crime.[6] Thus, while the Caribbean is frequently portrayed as an exporter in the flow of illicit goods, the region is in fact also a massive importer—of smuggled cash.

The same modes of transportation used to smuggle drugs into the United States are likewise used to smuggle cash from retail sales back out of the United States to pay wholesalers and sustain the larger criminal enterprise.[7] While some cash therefore returns to traffickers using many of the maritime methods discussed earlier, most interdicted shipments occur in airports. The Department of Homeland Security suspects this trend may simply be a result of the enhanced screening available for air travel when compared to land and sea conveyance (especially small maritime vessels).[8] And while money laundering is increasingly trending digital, the all-cash business of illicit smuggling continues to necessitate the movement of bulk shipments of cash. These shipments, running parallel and in reverse to trafficking flowing through the Caribbean transit zone, employ an equally diverse range of techniques. Simpler efforts include putting cash-stuffed envelopes in the mail or employing couriers, known as "smurfs," to carry small amounts. More sophisticated enterprises include "complex laundering operations involving front companies, offshore banks, and correspondents and intermediaries in multiple countries."[9] Law enforcement has interdicted Colombian-bound cash transiting the Caribbean, "secreted into cars, dolls, television sets" and even in "shipments of bull semen" (seriously).[10] Remittance networks, which I referenced earlier, are yet another means of facilitating the movement of money throughout the region and have been shown (as Young notes) to exacerbate money laundering and related crimes.[11]

Once safely back in the hands of trafficking ringleaders, smuggled cash is used for a variety of applications. Some is used to pay bribes to keep networks open, some is reinvested in purchasing cocaine from wholesalers, and some is laundered by laundering professionals for fees upward of fifteen cents on the dollar.[12] An unknown quantity may also be destined for terror financing, as I explore in the previous chapter. Former Drug Enforcement Administration (DEA) administrator Asa Hutchinson notes that there is a "strong case to be made that drug trafficking proceeds are being funneled to terrorist organizations," the potential for which was made clear in DEA assistant administrator Steven Casteel's remarks on the relationship between cocaine distributors and Afghan narcotics producers.[13] (Such comments, while not insincere, also likely reflect a perceived need in the aftermath of 9/11 for federal agencies to remain relevant and secure budgets.) Finally, while bulk cash payloads frequently make their first stop in Mexico, DHS trends from airport seizures suggest that most illegally transported money is ultimately destined for points farther afield, likely moving by maritime means at some stage.

The scale of bulk cash being smuggled across the transit zone is difficult to imagine (earlier I relayed an estimate that laundering of Caribbean proceeds from transnational crime exceeds $60 billion annually). In fact, so much money crosses between the United States and the transit zone that it is difficult to launder all of it—and so it begins to pile up. One Chinese-Mexican businessman, allegedly involved in the supply of precursor chemicals for the production of methamphetamine, was arrested in 2007 with $206 million in cash piled in his home (the largest single such seizure in history, and this from a midlevel trafficker).[14] The movement of cash intended for laundering, corruption, and investment in expanding transnational crime has even produced several models for bulk cash smuggling. The Department of Homeland Security categorizes these into two headings, the insource model (in which shipments are handled internally) and the outsource model (in which cash smuggling is contracted out). The former has obvious advantages in maintaining control over large sums of money while the latter ensures greater diversity of smuggling routes and allows transnational criminal networks to employ local trafficking infrastructure.[15]

The United States does provide technical assistance to help Caribbean authorities tackle the complexities of modern-day money laundering. The State Department's *International Narcotics Control Strategy Report* records that, alongside the Association of Supervisors of Banks of the Americas, the Federal Deposit Insurance Corporation has provided anti–money laundering seminars in the Caribbean, serving representatives from states such as Belize, Costa Rica, El Salvador, Honduras, Mexico, Nicaragua, Paraguay, and the Dominican Republic. Nevertheless, the unobserved linkages between money laundering, transnational crime, and even terrorist financing may remain extensive and, in the State Department's estimation, "only exacerbate the challenges faced by the financial, law enforcement, supervisory, and intelligence communities."[16] Money laundering is almost the quintessential multidimensional security issue; it acts as a lubricant for the diversity of Caribbean crime.

While the presence of transnational crime places the Caribbean at heightened risk of money laundering, it is the permissive attitude toward financial services that has served to propel the problem into an endemic culture. As a consequence, anti–money laundering is not accomplished strictly by transferring technical knowledge but also requires the recalibration of regional perspectives on the value of addressing money laundering. Thus, in so many ways, the issue of money laundering exemplifies the challenges discussed throughout this section. Money laundering is the quintessential multidimensional issue, intricately interwoven with all other criminal activities. Its prevalence is further aggravated by an environment that invokes a sense of complacency because a culture of movement and financial secrecy are, reasonably, valuable commodities in the resource-strapped Caribbean. Finally, money laundering's transnational connections typify the challenges facing maritime security. When it comes to money laundering—as with drug smuggling, irregular migration, terrorist financing, and gunrunning—"no individual country or single region stands alone—each is connected by a trail of dirty dollars."[17]

Misgovernance
Corruption and money laundering, like all of the transnational issues discussed here, are interconnected. The extensive infusion of corruption and graft in regional states ensures that the Caribbean's offshore banking services remain "ideal" for the laundering of criminal money, according to law professor Mary Alice Young.[18] Thus, through corruption, Caribbean states are at risk of becoming facilitators of gray- and black-market economies that encourage the development of still more illicit ventures. The implications of corruption are potentially volatile. This is on full display across feral and undergoverned communities, as I introduced in chapter 1. Garrison districts in Jamaica offer one such example of what happens in the fusion of governance and crime. In these communities, underpaid law enforcement are vulnerable to bribery by local dons, "the de-facto leaders of these areas," who use a mix of violence and persuasion to keep neighborhoods aligned with a specific political party. The result is that "authorities are essentially providing ungoverned functional spaces for the illegal gangs to operate" in exchange for a reliable base of support.[19]

In 2010 the Jamaican government relented to American pressure to arrest the head of one such district, Christopher Coke, the leader of the Shower Posse syndicate. Coke, as David Kilcullen chronicles, reigned in the Kingston community of Tivoli Gardens as the de facto mayor, delivering political support for his patrons and administering a wealth of extralegal social and judicial services in his fiefdom.[20] However, Jamaican efforts to catch Coke epitomized the pitfalls of urbicide I introduce in chapter 1. The community of Tivoli Gardens was besieged by police and the military, and several community members were killed as they defended their neighborhood. The tactics employed by Jamaican forces typify the shortcomings of "cutting" military strategies and "pasting" them in police scenarios. A cable sent by the U.S. Embassy's chargé d'affaires Isaiah Parnell is worth quoting at length:

Assault on Tivoli Gardens Stronghold

Following two days of civil unrest and gang-related violence in several parts of the metropolitan Kingston area, at midday on May 24 the [Jamaica] Defence Force (JDF) launched an all-out assault on the heavily-defended Tivoli Gardens "Garrison" stronghold controlled by Christopher "Dudas" Coke, the alleged overlord of the "Shower Posse" international crime syndicate who is wanted to face extradition to the USA on drugs and weapons trafficking charges.... The JDF fired mortars and then used bulldozers to break through heavy barricades which Coke's supporters had erected to block entry to the fortified enclave. As of 6:00 p.m. May 24, heavy fighting continued within Tivoli Gardens, and a fire was burning out of control in the adjacent Coronation Market. The JDF plans to continue operations through the night. Large numbers of women and children have fled the area. While casualty figures are not yet available, at least six JCF/JDF officers are known to have been killed in recent days, and many more wounded; the numbers of gang members and civilians killed or wounded are not yet known.[21]

The cable goes on to detail ensuing gang violence across the city, insufficient hospital resources, and flight cancellations through the next morning. In the aftermath of Operation Garden Parish, as Kilcullen illustrates, the scene was surreal: "Kingston had become a war zone in the course of enforcing a United States extradition request against a single international drug trafficker."[22] Tivoli Gardens exemplifies how, to borrow a description from Kimberley Thachuk (formerly of the George Washington University), governance failures have precipitated the "perfect ecosystems for private power brokers" as well as the unique challenges in dismantling them (Coke eluded capture for weeks).[23] Thachuk continues on to note that, in the absence of order, and when government falls victim to rampant patronage, drug lords have the capacity not only to govern real political space but also to captivate the genuine admiration and loyalty of large swaths of the population.

Police and government officials are on the "payroll" in many Caribbean communities, as in Tivoli Gardens. The result is a capsizing of public confidence and a sustained flow of narcotics, weapons, and people throughout the region. Anecdotes on corruption—which, like all things illicit, tell only a fraction of the story—suggest an extensive system of criminal infiltration. In 1991 Trinidad and Tobago's assistant commissioner of police Rodwell Murray alleged that a drug-trafficking cartel was operating within the police department.[24] A later report expanded on Trinidad's intricate system of corruption in which competing factions hidden within the police force recruited officers from early in their careers. These mafias provided career services, including promotions within the force, protected members, and opened access to sources of patronage.[25] A nonrepresentative sampling of informants in an International Organization for Migration study from the Bahamas, Jamaica, St. Lucia, the Netherland Antilles, and Suriname all believed that local immigration officers either facilitated or ignored instances of human trafficking in their country. In Barbados, a passport scam documented by the International Organization for Migration implicated law enforcement, who were accused of facilitating the passage of irregular migrants from Guyana. Trinidad and Tobago police have also been arrested in connection with ransom kidnappings.[26] In the summer of 2013, the United States withdrew aid from St. Lucian law enforcement after allegations that police had participated in extrajudicial killings.[27] Ironically, the killings coincided with St. Lucia's Operation Restore Confidence, an initiative to battle rising local crime rates that had been negatively impacting tourism.

It is not just poorly paid beat cops who are involved in transnational crime. In August 2013 the director of Suriname's counterterrorism unit, Dino Bouterse, was arrested in Panama and turned over to the DEA. He was accused of narcotics trafficking, with an indictment alleging that Bouterse smuggled cocaine in a suitcase on a commercial flight across the Caribbean that July.[28] The Associated Press reported that Bouterse had previously been sentenced to eight years in prison in August 2005 for fencing cocaine, weapons, and stolen cars. Even more astounding than a country's counterterrorism chief being arrested on multiple occasions for narcotics trafficking, gunrunning, and smuggling, Bouterse's father (himself an alleged drug trafficker, and in fact convicted of such in absentia) became president of Suriname in 2010. And corruption runs right to American shores. In May 2013 the director of environmental enforcement in the Department of Planning and Natural Resources in the U.S. Virgin Islands was arrested on drug-trafficking charges after allegedly using a government boat to transport cocaine. Then-DEA administrator Michele Leonhart expressed concern over the rise in drug trafficking in the Caribbean and its propensity to corrupt law enforcement in American Caribbean territories.[29]

The impact of this corruption on communities is difficult to measure but undeniably degrading. Anthony Maingot notes of St. Kitts and Nevis, for example, that corruption eroded confidence in the government and that basic governance and sovereignty suffered as a consequence. Corruption acts as an echo chamber, amplifying the psychological context of the Caribbean security environment. Perceptions of governmental fragility have the potential to become so pervasive that at one point Sir Shridath Ramphal, former commonwealth secretary-general, noted without exaggeration, "It only takes 12 men in a boat to put some of these governments out of business."[30] Narcotics trafficking is unlike any other criminal enterprise in its capacity to spread corruption and graft, which has profound and often understated impacts throughout domestic politics and international relations. The interplay between corruption, context, and disorder is no more clearly exhibited than in the confusion over the identities of those committing crimes. In many places murders are sometimes performed by men dressed as police, and it is far from obvious whether the assailants are "thugs masquerading as policemen" or genuine officers moonlighting for cartels. Meanwhile, when the government makes a large bust, law enforcement officers "pose for photos . . . brandishing assault weapons, their faces shrouded in ski masks, to shield their identities. In the trippy semiotics of the drug war, the cops dress like bandits, and the bandits dress like cops."[31]

Obviously, corrupt agents of law enforcement and national security hamper the effective action of those organizations. But the implications of corruption reach far deeper. Poor or lacking governance, as I explore in chapter 1, invites vigilante, criminal, and terrorist groups to fill the ensuing power vacuum. And critically, as the broken windows theory suggests, it is not only the inefficiency of corrupt authorities that invites competition from nonstate actors; even the very perception of corruption can breed disorder. Vigilantism took root so strongly in the Caribbean between the 1980s and the early 2000s that, at one instance, the national security minister for Trinidad and Tobago was compelled to issue a notice on national radio and television as a reminder that self-styled self-defense organizations remained illegal.[32] There are also political ramifications for reduced confidence in the government. Public disaffection undermines the validity and stability of the democratic process, which puts the very authority of the state in jeopardy.

Finally, not only does corruption degrade perceptions of government, it impacts the capacity of the government to defend its own survival. In chapter 4, I explore the debilitating implications for communities of the illegal trade in weapons. When law enforcement is complicit in gunrunning, the devastation reaches even deeper as authorities facilitate their own reduced capacity to respond to violence.[33] Similarly deleterious impacts can be seen from human trafficking. The corruption of border personnel, law enforcement, and politicians directly impedes the adequate enforcement of existing laws criminalizing human trafficking (however lacking they may be). This leaves communities vulnerable to greater human trafficking, which, as I highlighted in chapter 4, is part of an even wider network of irregular migration. Every tentacle of illicit transnational crime inevitably reaches to the heart of Caribbean states through corruption and graft. With a hold on the very institutions designed to safeguard the people, crime and trafficking can suffuse the body politic with corrosive ease.

LESSONS FROM THE COAST GUARD

Transnational issues (which predominate in the littorals) are among the defining problems of a globalizing world. No branch of the American armed forces perceives this better than the U.S. Coast Guard, which is dedicated to combating transnational crime. We can glean some important lessons from the Coast Guard's experience when considering a broader maritime security strategy.

The Coast Guard is aided in its transnational security pursuit by its decidedly hybridized nature; it is, as stated in the force's 2014 *Western Hemisphere Strategy*, "the only military law enforcement organization which is [also] a member of the national intelligence community," and thus "bridges traditional enforcement gaps between military and law enforcement organizations."[34] The Coast Guard's inherent cross-bureaucratic nature seemingly frees the organization to conceive of its mission in different (perhaps broader) strategic terms than that of other maritime services. This conceptualization was on display during the Vietnam War, when the Coast Guard applied its countersmuggling expertise in Operation Market Time, a U.S. Navy–led initiative to counteract resupply of communist guerillas from the sea. This institutional flexibility is likewise on display on America's northern border with Canada. Integrated cross-border maritime law enforcement operations are specialized missions that permit certain U.S. Coast Guard and Canadian officers to hold law enforcement authority on both sides of the border. This is a highly noteworthy act for an American military branch, ceding a degree of sovereignty, designed to undercut one of transnational trafficking's greatest advantages (borders).

This strategic flexibility is mirrored across the Western Hemisphere Strategy. Of the three primary "priorities" identified by the Coast Guard in this document, the first is explicitly intended to address the spread of transnational networks, which present "new threats to maritime safety, security, and efficiency." The Caribbean Basin, Central America, and Mexico all face the "destabilizing impacts of violence, corruption, terrorism, natural disasters, and trafficking in drugs, humans, and arms." The Coast Guard's recognition of the ambiguity and flexibility of these networks is particularly notable and reflective of the multidimensionality explored above. The authors relay that organizations that "evolve and mature for one illicit purpose have shown an increasing propensity to diversify their nefarious activities." They continue, "In the next decade, the links between networks initially formed for illicit activities including drug smuggling, human trafficking, or terrorist activity will continue to blur."[35] In response, the Coast Guard has elucidated a strategy with some similarities to a broken windows approach. An exploration of this strategy shows the feasibility of such an approach for maritime authorities and provides greater theoretical grounding to the strategy by assessing it in relation to the broken windows theory.

A significant evolution in the Coast Guard's approach to Caribbean security is its emphasis on networks. Recognizing the perennial shortfall of resources required to make interdiction a successful deterrence strategy, the U.S. Coast Guard's paper instead turns to dismantling transnational criminal organizations themselves. This approach moves pointedly away from the midcentury police enforcement model implicitly rejected by George Kelling, James Wilson, and Catherine Coles. That model, you will recall, is both resource intensive and of limited success in catching criminals in the act. The Coast Guard's approach to countering networks consists of three prongs: understanding and fostering a network culture of its own, identifying networks, and prosecuting networks.[36] Of these, the greatest challenge to a hierarchical organization comes in understanding the fundamental components of networks and co-opting their most dynamic qualities. Here our study of broken windows can provide some direct insight for the Coast Guard.

Police commissioner William Bratton achieved such a feat in New York City, pushing decision making down to precincts and empowering communities to help set the enforcement agenda. The application of community policing necessitated a less centralized understanding of neighborhood needs. This bred a federalization of enforcement, offering officers on the streets and their immediate superiors greater flexibility to make operational decisions in real time. Similar to the recommendation of counterinsurgency theorists such as John Nagl and David Kilcullen to push decisions down the chain, Bratton's disaggregation of command produced responses better suited for local needs and better equipped to leverage time-sensitive intelligence. In an effort to meet community demands, broken windows policing produced a network of enforcement in place of a strict corporate system. Better collection and timely employment of intelligence, stronger community relationships, and stronger institutional effectiveness—all components of the Coast Guard's emphasis on defeating networks—are also proven outcomes of the broken windows theory, as seen in chapter 2. These lessons can directly guide the Coast Guard's efforts to embrace network culture, which it outlines in its strategy as "decentralized authority, highly adaptable members and practices, valuing competency above rank, perpetual self-analysis, shared 'consciousness,' developing partnerships with non-traditional actors, and achieving a national 'unity of effort.'"[37]

In conversation with Rear Adm. Charles Michel at a conference on the sea services in Washington, D.C., in 2014, the then-deputy commandant for operations remarked to me the ethos that no Coast Guard asset should turn a propeller without a purpose. The admiral was referring not only to the more economically efficient use of assets but also to the smarter use of ships and aircraft to defeat networks instead of more labor intensive and less effective patrols. Admiral Michel placed particular emphasis on the *Western Hemisphere Strategy*'s pivot to networks, both combating them and mimicking their construction. Admiral Michel recognized the challenges in adapting a bureaucratic military organization to better reflect a network. His evident appreciation for the difficult institutional transformation this requires is indicative of the U.S. Coast Guard's understanding of the complexities of, and the importance of combating, hybrid maritime threats. In a subsequent interview, the admiral even made explicit reference to the Coast Guard's aim to produce a "tipping point" on crime in the Caribbean, reminiscent of Gladwell and Kelling's assessment of broken windows' tipping point influence on crime in New York.

Even the more traditional components of the *Western Hemisphere Strategy* highlight an important rhetorical shift, one closer to community policing and the broken windows theory. The document suggests strengthening interdiction capacity by expanding the scope of suspicion beyond narcotics trafficking to include any kind of transnational crime. This recommendation would provide the legal and strategic recognition of one of broken window's primary premises: that violent crime is not atomized, not a stream of isolated incidents. In theory, broken windows expresses this interconnectedness in the process I outline in chapter 2 (broken windows lead to abandoned houses, which lead to drug use, which leads to panhandling, which leads to muggings etc.). In practice, broken windows also expresses this interconnectedness by demonstrating that policing quality-of-life crimes puts police in contact with wanted felons and knowledgeable informants (e.g., about one fare-beating arrest in ten on the subway produces someone wanted for a Class A misdemeanor or felony). By focusing on the climate of enforcement instead of specific crimes (e.g., narcotics)—which is consistent with the Coast Guard's newly expressed perspective on networks—broken windows would suggest that maritime enforcement bodies stand a greater chance of impacting Caribbean security. While such implementation lags, the inclusion of language suggestive of a broken windows lens lends greater credence to the argument that the mechanisms identified in community policing are likewise applicable in the maritime domain.

The document also provides several suggestions that could directly impact the context of Caribbean security, the most significant element of any broken windows approach. For example, the strategy recommends expanding the deployment of LEDETs (law enforcement detachments) to "non-traditional platforms and/or locations with increasing TOC network activity and threats."[38] This initiative would not only increase the reach of American enforcement efforts but would place U.S. authorities in direct contact with local authorities and communities. In so doing, LEDETs become sounding boards for Caribbean communities in much the same way many officers do in neighborhoods policed by a community-oriented approach. Gathering intelligence is a positive outcome of this contact, but the primary assets are in expressing an honest commitment to improve a community and in genuinely understanding and responding to local needs. As a result, LEDETs and their partners can help restore an element of self-efficacy among Caribbean communities, which would magnify their efforts more than any new cutter or warship could. For this same reason, Southern Command's decision to cancel civil affairs deployments in fiscal year 2013 under sequester budget pressure should be seen as contrary to the community policing value of building local familiarity and credibility.[39] Another Coast Guard component of shaping the context of Caribbean security is in providing for an effective, "persistent presence" in the Basin. Such a presence, alongside gathering timely intelligence, would similarly signal the enduring commitment of the United States (and, critically, its local partners) to the mission of bettering life for local communities. To this end, the U.S. Coast Guard, with its smaller ships and valued law enforcement reputation, is thought of within the United States government as a highly sought-after military partner for Caribbean states.[40] This is so in part because, for many Caribbean states, constabulary and police bodies maintain the primary obligation for law enforcement and security policy, with the military serving as an auxiliary.[41] As such, the Coast Guard is a more relatable institution. Similarly, many of the security contributions from major Western states come in the form of constabulary or police forces, notably Scotland Yard and the Royal Canadian Mounted Police.[42]

Complementing the Coast Guard's law enforcement role is the force's interest in enabling access to the littorals, which has the potential to bring the U.S. Coast Guard, LEDETs, local partners, and local communities into even greater and more productive contact. And while the Navy has not always been able to deploy goodwill missions in the Caribbean Basin over the past few years, the Coast Guard's partial role as a relief organization has left it well-adapted for humanitarian missions in the hemisphere. After the 2010 Haitian earthquake, the U.S. Coast Guard was the first American agency with assets on hand (including almost one thousand Coast Guard service members).[43] A broken windows perspective on maritime security would suggest that the goodwill generated by humanitarian aid and disaster relief is as important to regional security as are conventional enforcement operations. Indeed, the infrastructure support provided by humanitarian aid and disaster relief has the capacity to literally fix broken windows, restoring a sense of order in communities hit hardest by crime or nature. The argument here is not, however, that the Navy cannot play a productive maritime security role in the Caribbean or elsewhere. Rather, recall the debate on the littoral combat ship from chapter 1—operating in the littorals requires investments not only in strategy and tactics but also in hardware. The Navy can learn from the Coast Guard in many of those respects.

There are, however, areas in which the Coast Guard's strategy could learn from our exploration of broken windows. Indeed, targeting networks, as the strategy poses, is not a flawless approach. As we have encountered throughout this section, the underground world of illicit smuggling and trafficking is notoriously difficult to understand because of a dearth of available data. Maingot wrote of this void of knowledge, even in the basic relationship between the primary syndicates of the time: "Are the three reputed *mafias* in Trinidad and Tobago . . . or the Jamaican posses, or the

Dominican *mafias*, subordinate to the Cali cartel? Or are they autonomous players that act sequentially, segmentally, and local in a chain of individual conspiracies running from producer to consumer?"[44] Perhaps the greatest such assumption the Coast Guard's new strategy makes is that with enough intelligence, these structures can be understood well enough to be dismantled. And surely the Coast Guard, as an intelligence agency, has access to sophisticated information flows. Yet the broken windows theory provides a means not just for collecting intelligence (as I just discussed) but for avoiding relying too centrally on intelligence. The Coast Guard should place even greater emphasis on impacting the environment in which all networks operate instead of investing too deeply in identifying and targeting particular organizations or individuals. Networks will replace one another, but context is what turns the tide.

Lessons Learned

Trafficking in the Caribbean is not new. The trade stems from a history spurred on by the region's earliest European colonizers. With the introduction of Spanish restrictions on trade and migration in the early sixteenth century, the Western Hemisphere was ensnared in a global network of smuggling. Just as today, the implementation of government regulations inevitably provided an arbitrage opportunity for individuals willing to shoulder the risk, known to Spanish colonialists as piratas (pirates), or what the British called the "private trade."[45] For centuries, this trade made use of ample lawless territory. Such ungoverned spaces "were out of control, but they were also out of touch, having little if any impact on the major concerns of central governments."[46] Out of sight and out of mind, the region slowly adapted what some have called a "culture of smuggling," one that "affects the very survival of many of the region's governments."[47] Over the past century, as signaled in Kilcullen's megatrends from the first chapter, interconnectedness has replaced isolation, and this culture of smuggling has burst forth as a genuine threat to Caribbean and international security.

I have noted that financial and political constraints across the Caribbean serve as obstacles to effective regional security apparatuses. At the expense of external patronage, Caribbean security initiatives frequently embody the strategic vision of the patron. What results is an important divergence, with a "multi-dimensional view on security threats predominating in the region" and a "more narrowly focused law enforcement emphasis" among U.S. officials.[48] This asymmetry strains efforts to build strong partnerships, which, in the words of General Kelly, are "the cornerstone of

U.S. Southern Command's engagement strategy and [are] essential for our national security."[49] This dedication to partnership is a key element of the maritime strategy for regional security. Indeed, in light of the profoundly transnational character of Caribbean crime, economy, and sociology, the Coast Guard and Navy have come to place considerable weight on their reliance on international partners.

Given the centrality of these partnerships to American capabilities and the effective maintenance of Caribbean security—recall that half of successful Joint Interagency Task Force South operations would not have taken place without international cooperation—any strategy the United States employs in the Caribbean would be markedly more successful if it were one of consensus. Yet unified action has been difficult to effect, and much of the issue is institutional. In the United States, dozens of federal agencies participate in counternarcotics missions, but "without clear lines of authority [and] the absence of a lead agency . . . the counter-narcotics struggle is in jeopardy of falling between the same regional bureaucratic gaps as have other unconventional threats."[50] Further, myriad intergovernmental organizations throughout the Basin struggle to command the resources and authority necessary to motivate sustained local action. These organizations are reflections of the geography that engendered them. The Caribbean Basin, according to the Coast Guard, is home to more than 1,200 islands under the jurisdiction of more than twenty-five countries. The Caribbean Community represents fifteen states, most of British tutelage, while many of the Central American states lining the Basin are represented both in Caribbean organizations as well as the more exclusive Central American Integration System. The Caribbean is noteworthy for having the greatest concentration of defined maritime borders on the planet, but even this accomplishment invites unique bureaucratic hurdles in fighting transnational issues.[51] Complicating factors even further is the high concentration of foreign dependencies, with local territories under the authority of the Netherlands, France, the United Kingdom, and the United States. With so much institutional fragmentation, "only an integrated strategy that synchronizes the various organizational efforts under a unifying concept can address these problems."[52] As I explore in chapter 4, the community policing framework, of which broken windows is a subset theory, is among the few security ideologies of consensus in the region. Constructing an American maritime security strategy in the language and approach of a community policing framework would better enable regional security cooperation by elevating American leadership beyond strictly the financial and technical to the strategic and ideological.

Multidimensionality and Context

General Kelly summarizes the leviathan of transnational crime as follows:

> Picture an interconnected system of arteries that traverse the entire Western Hemisphere, stretching across the Atlantic and Pacific, through the Caribbean, and up and down North, South, and Central America. Complex, sophisticated networks use this vast system of illicit pathways to move tons of drugs, thousands of people, and countless weapons into and out of the United States, Europe, and Africa with an efficiency, payload, and gross profit any global transportation company would envy. In return, billions of dollars flood back into the hands of these criminal enterprises, enabling the purchase of military-grade weapons, ammunition, and state-of-the-art technology to counter law enforcement. This profit also allows these groups to buy the support—or silence—of local communities through which these arteries flourish, spreading corruption and fear and undermining support for legitimate governments.[53]

Threats to maritime security in the Caribbean exhibit the ambiguities of the hybrid threats discussed in chapter 1. Some nonstate actors have even grown so large and influential that they are thought to leverage more financial and military weight than some regional states. The Coast Guard refers to this hybridization directly, referencing in its 2014 strategy a "crime-terror-insurgency nexus" that connects transnational crime to political violence across the Basin.[54] And while the Coast Guard is cognizant of the growing threat to the maritime domain posed by hybridized transnational threats, the force's strategy suggests that the deep impacts of illicit trafficking are only now making themselves fully understood. As we saw throughout chapter 3, the Coast Guard's more progressive concepts have yet to be implemented at scale in practice. Interdiction remains the most persistent form of American counternarcotics in the transit zone. And by some metrics, that strategy has had remarkable success. Between efforts in the Pacific and Caribbean, not a week passes without reports of astronomical drug seizures. Such an approach, however, has not systematically reduced the overall flow of drugs, which continues "virtually unabated."[55]

The problem is twofold. First, as I note throughout this section, such an approach deemphasizes the flexibility and multidimensionality of modern networks. Second, combating a culture is far different than combating a person, organization, or even a network. Naím captures both of these components writing of the Mexican kingpin Chapo Guzman. He presciently noted in 2006 that Guzman was destined to be killed or captured, but the victory would be short-lived. (Guzman was indeed captured, then dramatically escaped, and then was subsequently recaptured.) This rotating door of high-level crime bosses isn't surprising. The real problem, the one that defies easy solutions and is often even harder to discuss, is "the diffusion of the drug business into the fiber of local and global economic life.... More than any cartel, kingpin, or rebel warlord, it is this pervasive global mainstreaming of the business that the fight against drugs is up against."[56]

The nature of the threat has changed, and enforcement bodies have struggled to integrate these realities with the strategies they employ. As I noted in chapter 4, the militarization of crime has produced an impulse toward militarizing law enforcement, with mixed results at best. With Caribbean police occasionally overstretched in pursuit of law and order, it is not uncommon to have "armies deployed on the streets and on the seas alongside the police in support of anti-crime efforts."[57] While the added resources of the military can be invaluable, the strategic conceptions common among militaries make them ill-equipped to operate as auxiliary law enforcement. Turning the region into a "fortress Caribbean" is not only inimical to tourism but to the very survival of local communities (recall the idea of urbicide).[58] And as General Kelly notes, "Given its history, the region is sensitive to any appearance of increased militarization."[59] Until such time as police and the military can effectively combine the best of their strategies and resources, respectively, criminal groups will continue to outpace the enforcement strategies designed to combat them.

Context is always harder to directly observe than the multidimensional linkages of crime, but the concept is embedded in comments on "a spreading culture of lawlessness" and "no heritage of respect for the administration of justice."[60] This cultural ambiguity toward central authority and a marred criminal justice system is evident across the case study. It is evident in Alfonzo, the Mexican entrepreneur engaging in small-scale trafficking without regard for the rise and fall of kingpins. It is evident in the Costa Rican fishermen pressed to provide way stations and supplies for transiting traffickers. It is evident in the vast and poorly regulated torrent of human traffic, whisking victims across borders. It is evident in the smuggling of migrants throughout the Basin, many of whom become victims or facilitators of transnational crime in the honest pursuit of a better life. It is evident in the free flow of billions of dollars smuggled, laundered, and invested in drugs, guns, or campaigns of terror. It is evident in the rise of self-defense militias, fractured sovereignty, and the popularity of gang leaders like Christopher Coke. It is evident in the proliferation of local violence, tied directly and indirectly to transnational crime. And finally, it is evident in the graft and corruption that blocks meaningful progress on Caribbean security. Context pervades and defines our case study, yet its significance is underscored most effectively only if you are looking for it, through the lens provided by the broken windows theory.

In fact, references to context and communal self-efficacy (or lack thereof) predominate across the Caribbean literature. Historically, traffickers have repeatedly sought out ungoverned spaces and regions of "tolerant government attitudes" (Zackrison's words) in which to ply their craft.[61] The former Colombian ambassador to the Organization of American States, Fernando Cepeda Ulloa, made reference to the corrosive impact of this culture when he noted the region's "tolerance of criminal conduct" and an abiding sense of the "invincibility of the drug barons."[62] The ambassador also noted that this psychological defeat is intertwined with the consequences of the absence of any internationally supported counternarcotics strategy.[63] In testimony to the House Armed Services Committee, General Kelly made similar note of the region's "corrosive criminal violence," which is enabled in part by "permissive environments for illicit activities."[64] Two decades earlier, the West Indian Commission expressed dismay at the apparent "powerlessness" of regional governments in the face of transnational crime.[65] Zackrison refers to a "culture of smuggling that permeates the Caribbean."[66] In chapter 3, I quoted an Associated Press article that used the word "anxiety" to characterize the mood of a small Jamaican fishing village concerned about expanding trafficking. Even the U.S. government references perceptual obstacles to security in the region. In the State Department's 2014 International Narcotics Control Strategy Report, the authors cite a "culture of impunity" in Guatemala (language that resurfaces in the 2016 version) and the need for "culture of lawfulness training" in Mexico, the only two instances in the entire report where this wording is employed.[67] The *International Narcotics Control Strategy Report* also references "frustration" in Jamaica, among both officers and the public, over the country's struggle to meet its criminal justice obligations prosecuting corruption and narcotics, which extract "a significant social cost."[68]

The broken windows theory offers compelling evidence that environment impacts the decisions we make. I already noted this impact in New York City as well as in literature across psychology and medicine. Yet we need not rely exclusively on extrapolating the theory's applicability from criminology and public health. In the weeks after the attacks of 9/11, narcotics trafficking across the U.S.-Mexico border plummeted by 80 percent not because of any new program but because of the *"perception* that security had been increased."[69] In the Caribbean, drug smuggling remained constant, but authorities registered a similar decline in illegal immigration "because of the same perception."[70] There are examples on an even more local level. In November 2000 DEA and Colombian counterdrug officers uncovered the first credible evidence of Russian–Colombian collaboration on the manufacture of narco-subs. Authorities were tipped off by villagers in the small town of Facatativá who became suspicious of Russian newcomers.[71] Police succeeded because locals believed authorities could respond to their needs and, more critically, because locals were invested enough in their community to identify and report threats. This is an ideal, organic representation of self-efficacy and community policing. Context therefore not only offers a means of understanding crime but a medium through which to combat complex disorder instead of any one particular crime, criminal, or criminal syndicate. If local Caribbean communities provide the "arteries" (in General Kelly's words) through which these networks flow, then broken windows can play a prominent role in shaping these pathways to help detect and restrict the movement of illicit actors.

❖ ❖ ❖

The avenues for applying the broken windows theory as a framework for maritime security in the Basin are extensive. And given the nature of the myriad hybrid threats plaguing Caribbean littorals, the application of a criminological theory to international security proves a relatively straightforward conceptual leap. Smuggling is a crime, after all, and traffickers are criminals. Epidemics of crime in New York City or the Caribbean may express themselves through different global implications, but the principles of epidemics and environment as expressed by Gladwell, Kelling, Coles, Wilson, and Bratton endure. Until now, American-led enforcement in the region has taken a combative approach to local issues, sidestepping communities without effecting change on a large scale. And yet society's role in alleviating Caribbean insecurity is not expressly new. Among others, Zackrison reached that conclusion when he remarked that the only way to effectively combat smuggling is "when done in concert with society." For Zackrison, however, this was strictly hypothetical, noting that "history records no such coincidence to prove the hypothesis."[72] On a macro scale, at the international level, that may be true. On a micro scale, though, this is far from theoretical. In cities that employ various iterations of community-based policing, authorities have demonstrated that policing with society is as possible as (and more beneficial than) policing against it.

In this section, we turned to the central argument of this project: How can we use the broken windows theory as a unifying framework for a maritime security strategy? By way of response, I began by describing the multidimensional nature of Caribbean security. As a result, we saw that academics and politicians in the region have long since argued for an approach to security that mirrors this reality. We also saw a clear need for more constructive enforcement solutions. Police and militaries are already blending in function and practice, but a concerted framework for how to do so constructively, and how the United States can best support that effort where appropriate, is lacking. The experiences of Caribbean enforcement appeal to both of our major themes—multidimensionality and context specificity—speaking to the values of "an integrated approach to 'the conditions which create instabilities in societies'" and a "'multidimensional approach to hemispheric security.'"[73] Moreover, I noted that Caribbean authorities have an affinity for community policing, which makes the broken windows theory a good starting point in developing a consensus-based maritime enforcement model. Finally, I identified how context cuts throughout, offering a domain in which hybrid threats can be combated under a single rubric.

As we can see from this section, the broken windows theory indeed provides suitable scaffolding upon which to construct a theoretical model of maritime security. What remains, however, is one of the most significant components of any theoretical evaluation: generalizability. And so, in the remaining two case studies in the section to come, I turn to the job of demonstrating that the lessons gleaned here are applicable beyond the turquoise waters of the Caribbean Basin.

PART THREE
Integrating Piracy

Nigerian military forces conduct bilateral visit, board, search, and seizure training on board the Military Sealift Command joint high-speed vessel USNS *Spearhead* (T-EPF 1) during the maritime security exercise Obangame Express 2015. *Spearhead* is on a deployment to the U.S. Sixth Fleet area of responsibility to support the international collaborative capacity-building program Africa Partnership Station. *U.S. Navy photo by Mass Communication Specialist 2nd Class Kenan O'Connor/Released*

CHAPTER SIX
Piracy and Perception in the Gulf of Guinea

In part 2, we took a deep dive into the Caribbean Basin to see whether key themes from the broken windows theory enable us to think about maritime security in a new and cohesive way. In so doing, I illustrated that maritime insecurity in the Basin could indeed benefit from a framing that focuses on multidimensionality and contextualization. In this section, the aim is to assess whether success in the Caribbean is an aberration or if a broken windows lens is truly generalizable as a tool for thinking about the littorals. To meet that final challenge, this section contains two regional illustrative cases, the Gulf of Guinea and the Straits of Malacca and Singapore.

First: West Africa. In the first half of this chapter, I illustrate, through a brief survey of the insecurities plaguing the Gulf of Guinea, the similarities these issues share with those in the Caribbean. By demonstrating that key features are shared between maritime disorder in the Caribbean and West Africa, we can see in a relatively short space how a strategic lens in the former region might be applicable in the latter. This portion of the chapter is broken into two sections. First, I detail the parallel nature of the narcotics trade in the Caribbean and the Gulf of Guinea, demonstrating clear relationships both in theory and in practice between the two. Second, I detail several of the quality-of-life crimes endemic to West African littorals. This discussion parallels that in part 2 (chapters 4 and 5) on the local crimes and context of Caribbean insecurity.

In the second half of this study, we encounter piracy for the first time. I demonstrate that piracy—despite being absent from the Caribbean discussion that served as our proof of concept—fits comfortably in the lens and language of broken windows. By illustrating that piracy can be integrated into the same narrative on maritime crime's multidimensional and contextual nature, we see that a broken windows lens is capable of adapting to new issues and locales and thus is not unique in its application to the Caribbean.

THE REGION

The literature is at pains to define the Gulf of Guinea's exact boundaries. Among narrower applications, the Gulf of Guinea is the body of water bounded by the prime meridian and the equator (0°0′0″, 0°0′0″), from around Togo to the northern portion of Gabon.[1] More expansively, the Gulf of Guinea is characterized as emanating from −15°0′0″, −15°0′0″, which would expand it as far west as Guinea and as far south as Angola.[2] Rear Adm. Adeniyi Adejimi Osinowo of the Nigerian Navy, who participated in the drafting of the African Union's Africa *Integrated Maritime Strategy 2050*, defines a "wider Gulf of Guinea" as the five thousand nautical miles of coastline from Cape Verde to Angola.[3] Ghanaian Navy officer commander Ali Kamal-Deen, on the other hand, employs in a *Naval War College Review* article an institutionally guided definition, using the membership of the Economic Community of West African States (ECOWAS) and the Economic Community of Central African States (ECCAS) as templates.[4] This would extend the region as far northwest as Senegal. The International Crisis Group (ICG) pursued a similarly institutionally guided definition using the Gulf of Guinea Commission, though this approach required ad hoc expansions—including Ghana, Togo, and Benin.[5] Freedom Onuoha, of Abuja's National Defence College, notes, in reference to this inconsistency, that the Gulf of Guinea is "almost too obtuse to be a gulf" at all.[6] Tension among these definitions subtlety permeates the literature, and, as Kamal-Deen observes, the definitions employed frequently "var[y] depending on the issue or interest at stake."[7] In line with our treatment of the Caribbean Basin, it is not a prerequisite to elevate one definition, although it is important to note in both cases that this geographic ambiguity is an important feature of the literature (and thus, sources of data).

As in the Caribbean, where local disorder in Colombia or Haiti can precipitate regional discord, so too is this the case in the Gulf of Guinea. Nigeria in particular is a linchpin, an economic and demographic powerhouse that for years has struggled to exert control over internal conflicts. That Nigeria shares direct sea borders with Benin, Cameroon, Ghana, Equatorial Guinea, and São Tomé and Príncipe serves to facilitate the "transnationalization" of these conflicts. The Niger Delta also exemplifies many of the risk factors explored in chapter 1. The delta is home to more than thirty million people, as many as ten million of whom are unemployed.[8] In the estimation of Francois Vreÿ of Stellenbosch University in Cape Town, the delta region lays claim to forty ethnic communities across just nine states, epitomizing the complexity and density we have discussed.[9] The topography is an extreme example of littorals with, as Vreÿ notes, a "vast transitional zone between the sea and the huge inland delta with its networks of waterways leading to and from the Gulf of Guinea." "These waterways," he continues, "serve as arteries for the syndicates" involved in the region's myriad criminal activities.[10] More broadly, Nigeria is a study in David Kilcullen's megatrends.[11] The country has the continent's largest economy and, approaching 200 million people, contains more than half of the West African population. Broader still, the region's coastal cities are, true to global trends, some of the most densely urbanized and highly populated parts of Africa. And it is these "dense conurbations along the coast," an International Crisis Group report notes, that "helped to create the conditions for an increase in crime."[12]

Maritime security has often taken a back seat to landward issues that are more immediately associated with threats to state sovereignty. This is starting to change, however. Raymond Gilpin, chair for defense economics at the Africa Center for Strategic Studies at the National Defense University, argues that the littorals are fast becoming a point of concern for governments, many of which are undergoing a "paradigm shift" in perception.[13] Unfortunately, in Gilpin's assessment, systemic underinvestment in maritime domain awareness as a result of that longstanding neglect of maritime issues has facilitated the use of West African littorals for illicit maneuver. Angola, with the longest regional coast at 1,600 kilometers, has a naval strength of 1,000 individuals—compared to the air force's 6,000 and the army's 100,000. Liberia, "the second-largest flag state in the world," has a coast guard of approximately fifty people and eight boats (all ten feet and under).[14] While security funding at-large may be weak, regional navies and coast guards have suffered particularly sustained neglect.

The Gulf of Guinea's rise in oil exportation has furthered interest in its littorals. As a link in the international energy market, the region is attracting ever-greater attention from abroad (in 2007, Nigeria overtook Saudi Arabia as the third-largest exporter of crude to the United States) and, as Onuoha argues, the very creation of U.S. Africa Command (AFRICOM) came in part from the economic strategic importance of West Africa for the United States (although the ever-widening war on terror was likely an important factor as well).[15] The boon in production has also been a windfall for militants, fueling local conflicts through the sale of illegally bunkered oil. West African production accounts for well over half of the African total. That crime often emanates from Nigeria is in part due to its significant share of that production (and thus access to militant financing), about half of the regional total. Like the Caribbean, the region is also at trade crossroads—it is placed at the center of trade lines that connect North America and Europe to West, Central, and South Africa—which makes local insecurity relevant to international actors. This geographic placement, combined with state weakness, has precipitated the use of the Gulf of Guinea as a transshipment point, similar to the Caribbean's role in transnational crime. Through recent history, the Gulf has attracted the Russian mafia, South American cartels, and Lebanese smugglers, among others.[16]

Alongside terrorism, maritime insecurity is now, in the estimation of the ICG, among the foremost security concerns for African states. In response, rhetoric in the Gulf increasingly reflects that which we saw in the Caribbean context; multilateralism and the multidimensionality of crime are emerging tropes. In an interview with *MilTech*, Rear Adm. Geoffrey Biekro, then chief of naval staff for the Ghanaian Navy, remarked that it was a "commonly shared picture that most of the criminal activities going on in the whole region are transnational" and that "the need for multilateral-cooperation cannot be over-emphasized."[17] Gilpin likewise writes that the Gulf's potential and wealth is being "undermined by multifaceted domestic, regional and international threats and vulnerabilities."[18] As in the Caribbean, such multifaceted insecurities include narcotics trafficking, weapons trafficking, vigilantism, irregular migration, the specter of terrorism, and corrosive corruption.

Narcotics

West Africa has emerged as a significant route in the trafficking of narcotics, particularly South American cocaine. As Mexican cartels grew in influence, as trafficking routes came under greater surveillance, and as demand grew in Europe, West Africa became a "hub" for the trade.[19] In 2007 the UN Office on Drugs and Crime (UNODC) estimated in a report on cocaine trafficking in West Africa that more than a quarter of the cocaine annually consumed in Europe passed through that region (at a wholesale value of $1.8 billion).[20] Between 2005 and 2007, the report notes, 33 tons of cocaine were seized en route from West Africa to Europe. Previously, annual African seizures across the whole continent rarely surpassed 1 ton, while in 2007 seizures were more than sixty times those of five years prior. A 2015 estimate suspects that 2,200 pounds of cocaine are flown into

Guinea-Bissau every day, with larger shipments arriving by sea.[21] The largest such seizures have also generally taken place on the high seas—the UNODC reports that five of the seven largest seizures in one year took place on the water, for example.

Stephen Ellis of the Free University in Amsterdam notes that shipments have been detected across Cape Verde, Senegal, Mauritania, Guinea-Bissau, Guinea, Liberia, Sierra Leone, Ghana, and Benin.[22] Some examples include one seizure by a French naval ship of a Togolese tugboat laden with two tons of cocaine from Venezuela; the 2002 Nigerian seizure of a Brazilian ship carrying sixty kilograms of cocaine docked at the Tin Can Island wharf; and a 2003 haul intercepted between Senegal and Spain.[23] Ellis records an arrest in 2004 in the Ghanaian port of Tema, which uncovered more than five hundred kilograms of cocaine, and the detection in 2006 of a ship off of the Canary Islands carrying almost four tons of cocaine. In January 2008 more than two tons of cocaine was found by the French Navy on board the *Blue Atlantic* as it passed the Liberian coast on its way to Nigeria.[24] Lt. Cdr. Stephen Anderson, while serving as executive officer of HMS *Dauntless*, referred (in a Chatham House report) to this avenue as "Highway 10" since much of the trafficking from the Caribbean heading east takes place around ten degrees north latitude.[25] Although not geographically direct, the use of the Gulf of Guinea for drug trafficking is reminiscent of (and functionally similar to) Haiti's use in the Caribbean as a trafficking way station.[26]

Trafficking methods employed are familiar from our Caribbean study, such as using modified mother ships (freighters or large fishing vessels) that are then unloaded onto smaller fishing vessels for transfer. This has been the case in Ghana, which the U.S. Department of State (in the *International Narcotics Control Strategy Report* [*INCSR*]) identifies as a transshipment point for cocaine as well as Afghan and Pakistani heroin. In response, Ghana formed a marine police unit in June 2014 to operate as a de facto coast guard, and the United States has provided maritime law enforcement training for the Ghanaian Navy, emphasizing littoral narcotics interdiction. Once imported, narcotics typically leave the Gulf in one of two ways. Larger shipments are generally sent by sea, landing most often in Spain (which leads Europe in cocaine seizures). In UNODC's assessment, it is likely that Latin American cartels retain tight control over bulk shipments, as they often did when employing nascent Mexican cartels that initially provided protection and logistics support without directly purchasing the drug. The 2009 discovery of precursor chemicals in the region suggests that some cartels may have moved entire portions of their operations to West Africa.[27] Ellis notes that some prominent traffickers from Latin America are also suspected to have moved to the region. Moreover, the notion that cartels may be integrating the Gulf of Guinea into a global distribution chain lends greater significance to the rise of narco-subs, which can reach the African coast from Colombia.[28]

The portion of cocaine that is sold wholesale to Nigerian networks represents a parallel distribution network, and an increasingly prominent one. While the maritime route is typically used for bulk methods (according to the UNODC and in line with seizure rates), Nigerian schemes typically employ drug mules (or "swallowers," so called for swallowing cocaine in condoms) flying out of Lagos into Europe to move smaller packages.[29] As with Mexican cartels, the expanded use of payment in kind by producers will likely broaden the exploitation of mules as well as local narcotics abuse; drug usage already seems to correlate to regional transshipment patterns.[30] As Nigerian travelers face greater scrutiny, local traffickers have also increasingly turned to Nigeria's porous borders, including seaports, as reported in the *INCSR*. Regardless of transportation method, the presence of South American cocaine in Africa marks not only an issue of local human security but an obstacle in the fight against narcotics in the Caribbean. Proceeds from African trafficking will make their way west and will be used in part to further finance the Western drug trade.[31]

We can learn more about West African narcotics trafficking using our Caribbean case for a more direct comparison. For example, Ellis intimates that preexisting networks in the Gulf of Guinea offered attractive, ready-made partners for South American traffickers, just as local infrastructure in Jamaica and St. Kitts and Nevis was co-opted by cartels. One American drug official claimed, as relayed by Ellis, that Nigerian networks are so sophisticated that drug syndicates from both Asia and Latin America seek them out. The fluid dynamic of Caribbean transnational criminal organizations is also reflected in the Gulf, pointedly described by Ellis as an "adhocracy." Some Nigerian groups, for example, employ middlemen who act as insulators between foot soldiers and bosses, ensuring that relationships are often in flux.[32] Another component of Nigerian transnational crime is also familiar from part 2, particularly our discussion of Dominican traffickers—diasporas. North America and Europe are home to large Nigerian immigrant communities. For example, Moisés Naím notes that in Chicago, Nigerians were historically heavily documented in the wholesale market for heroin.[33] And as RAND's Audra Grant writes, in an analysis of U.S. heroin seizures in 1998, the majority was connected to Nigerian networks.[34] Largely blameless, immigrant communities unfortunately served as a substrate within which traffickers navigated with obscurity, manipulating for their own ends the accountability and trust common in communal ethnic networks.

Narcotics trafficking in the Gulf of Guinea is also multidimensional. Ellis notes, for example, that the militants in the Niger Delta who steal (illegally bunker) oil from pipelines sell that contraband to tankers waiting offshore in exchange not just for cash but weapons and cocaine as well. As oil travels downstream, it represents one component of a complex regional market in which oil, drugs, and weapons serve as surrogates for cash. Narcotics may even play a role in funding extremism. A trans-Saharan smuggling route has been purported by some to provide funding for Al-Qaeda in the Islamic Maghreb.[35] As Andre Le Sage of the National Defense University's Institute for National Strategic Studies notes, this appears to be the case in Northern Mali, where Al-Qaeda in the Islamic Maghreb (alongside local militias) has operated protection schemes against traffickers.[36] A 2017 UNODC report similarly notes the involvement of both Boko Haram and Al-Qaeda in cocaine trafficking in the region.[37] While Vreÿ categorizes the volume of drugs entering from the sea as "alarming," he is quick to note that it is these multidimensional connections that are the most "disturbing" element of the trade, "dovetailing with criminal statecraft, terrorist groups and oil theft syndicates."[38]

Drugs have a long relationship with power and politics in West Africa, just as in the Caribbean. Former Liberian presidents Samuel Doe and Charles Taylor both did little to dissuade the use of Liberia as a staging ground for trafficking, while former Togolese president Gnassingbé Eyadéma stands accused of complicity in narcotics trafficking.[39] In 1998, as Ellis chronicles, a former Nigerian senator was even arrested while in New York on charges of heroin trafficking. In 2005 American authorities likewise arrested a Ghanaian member of parliament carrying more than one hundred pounds of heroin.[40] Drugs have been used not only to enrich warlords and as goods to barter for weapons but also as a means of controlling pressganged militiamen. As Grant notes, drug use by combatants in Sierra Leone and Liberia was likely a contributing factor to the duration of the conflicts in those countries.[41]

Similar to misgovernance in the Caribbean, "criminal statecraft at times typifies governance" in the Gulf of Guinea.[42] This is nowhere more visible than in the case of Guinea-Bissau, which poses an interesting parallel to some Caribbean states. Like them, Guinea-Bissau is "off of most people's radar screens," in the words of a UNODC report. The country is "poor, weak, and yet not so unstable as to attract attention."[43] It has consequently been subsumed by narcotics. As with some Caribbean states, the value of transshipped cocaine in Guinea-Bissau likely exceeds the country's entire gross domestic product (GDP). The heads of United Nations peace missions gathered in West Africa have even expressed their concerns that Guinea-Bissau posed a threat to regional stability. Here, as in neighboring Guinea, Latin American dealers exercised some of their tightest control over the African–European route.[44] Le Sage notes that the country's narcotics problem earned it the distinction as the continent's first "narco-state." For illustration, he recounts that on just one day in 2009, both the president and a senior military official were killed in "tit-for-tat" assassinations linked to the trade.[45] Ellis relays the assessment of a Nigerian drug enforcement official, who alleged that Guinea-Bissau's military has provided warehousing space for outward-bound cocaine. At least one Revolutionary Armed Forces of Colombia (FARC) affiliate was arrested in the country in 2007, an obvious sign of the expanding connectedness of hybrid threats. FARC officials have also been reported in Accra, Conakry and Monrovia as well as other major cities in the region.[46]

Narcotics trafficking is not new to the Gulf of Guinea (Ellis succinctly summarizes the modern history of West Africa's involvement in the trade in a 2009 article in *African Affairs*). It is only over the past fifteen years, however, in Ellis' judgment, that bulk shipments have been consistently recorded entering from South America. As a consequence, a number of international actors now recognize the narcotics trafficking threat. Several West African UN peace missions, for example, have long since expressed the criticality of addressing both narcotics trafficking and connected phenomena of organized crime. The European Union's fact sheet on a strategy for the Gulf of Guinea references narcotics' billion-dollar industry.[47] Members of the Economic Community of West African States have already expressed the significance of cooperating on counternarcotics. The U.S. State Department's *INCSR* captures some of the ways the United States works to facilitate and support that cooperation at sea. AFRICOM, for example, lends increasing maritime law enforcement training in the region, including littoral interdiction training. The U.S. Coast Guard, in testament to the hybrid nature of Gulf of Guinea security, has participated in AFRICOM training, principally through the African Maritime Law Enforcement Partnership. The West Africa Cooperative Security Initiative, launched by the United States in 2011, also aims to address drug trafficking in the region. The United States has also provided vessels for littoral operations, including two ships donated in May 2012 to Nigeria's marine police.[48] Many such ships are former Coast Guard vessels. This was the case with NNS *Thunder* (F90), the former USCGC *Chase* (WHEC 721), acquired by the Nigerian Navy in 2011, as well as USCGC *Gallatin* (WHEC 721), which was transferred in 2014 and renamed NNS *Okpabana* (F93).

Given a wide variety of security and governance concerns in the region, both onshore and on the seas, the West African trade in narcotics is infrequently portrayed as the region's most pressing transnational issue, even as the *INCSR* has identified West Africa as a region of "growing concern."[49] Consequently, West African authorities have less counternarcotics experience and fewer antitrafficking resources than their Caribbean counterparts. Yet to dismiss West African narcotics trafficking as less than a relevant threat is to disregard our exploration of the broken windows framework's two critical lessons: multidimensionality and context. Geoffrey Till, using language reminiscent of broken windows, speaks to this dynamism: "Disorder at sea only makes things worse ashore. The success of transnational crime such as drug smuggling elevates the power of the kind of people who challenge civilized states ... it undermines their prosperity, security and ability to connect with other countries."[50] Such countries, he continues, tend to produce security challenges for those around them. Former president Barack Obama similarly warned against the regional "destabilizing effects of increasing drug trafficking in West Africa."[51] Like a vacant lot leading to loitering, fear, community abandonment, and crime, maritime disorder is contagious. Thus, while the region may face more prominent threats, this should not diminish the significance of the Gulf's emergence as a major hub in international narcotics trafficking. And just as in the Caribbean, this trade lends itself to portrayal through the broken windows theory.

On June 27, 2007, fishermen from the Senegalese village of M'bour noticed a strange boat, which they reported to the local gendarmerie, who found it laden with bricks of cocaine.[52] As with the Colombian villagers of Facatativá from chapter 5, who reported suspicious foreigners whom police found to be building a narco-submarine, this incident reflects just how critical police-community partnerships are, even in transnational crimes. Moreover, the converse is equally true. As the UNODC points out, "In a vicious cycle, citizen cooperation declines with each police failure, further undermining the ability of officials to do their jobs."[53]

The very rhetoric used to describe West African narcotics trafficking is reminiscent of the importance of perception and context we saw in chapter 5. Ellis, for example, notes that smuggling is "widely tolerated" and that "tolerance" from government officials encourages criminality. Ellis also refers to the significance of perception, noting that the expansion of narcotics trafficking in the Gulf of Guinea is a partial result of the "exceptionally favourable political context offered by ineffective policing" and "governments that have a reputation for venality."[54] Speaking of Ghana, the Department of State comments in the *INCSR* on the degrading impact of public perception and "rumors" of government complicity in trafficking. Of Nigeria, the department notes that the "perception of high levels of corruption and impunity encourages narcotic trafficking."[55] According to a former head of the Nigerian Drug Law Enforcement Agency, some of the agents charged with combating narcotics trafficking are even more invested in the trade than traffickers themselves.[56] Corruption, spurred on by enormous profits, entrenches criminal interests while "deepening fear and mistrust," characteristics seen in chapter 2 to be subtly destructive for communities.[57] Transparency International's Corruption Perception Index, whose attention to the sentiments of corruption is evocative of broken windows, ranked Nigeria 136th out of the survey's 176 countries in 2016.[58]

Colombian economist Francisco Thoumi affirms that illicit trades require "weak social and state controls on individual behavior, that is, a society where government laws are easily evaded and social norms tolerate such evasion."[59] The parallels between drug trafficking in the Caribbean and Gulf of Guinea suggest that this norm setting, such a critical element of the broken windows theory, is equally relevant in the West African context. Combating transnational crime in Africa will undoubtedly require fundamental investments in capacity building and situational awareness—even basic office supplies, according to the United Nations. Just as in the Caribbean, mere strategizing cannot overcome the limits of poor, weak states without enough fuel to power their police cars or enough ships to patrol their coasts. And yet, as we saw in part 2, by placing narcotics trafficking within the framework of a wider transnational flow, we can demonstrate the applicability of the broken windows theory's core principles—multidimensionality and context specificity.

Life and Crime

As in the Caribbean Basin, crime in the Gulf of Guinea is interwoven with national and transnational themes. This can be seen in an array of issues, from illegal fishing to smuggling to oil theft. These crimes are interrelated, often using the littorals as a shared domain for exchanging illicit commodities. The resultant security crisis has bred the proliferation of vigilantes and militias, which present yet more obstacles with similarities to the discussions on vigilantism in the previous section. I briefly highlight several such issues in the following section to explain the context in which security and disorder are framed in the Gulf of Guinea and to demonstrate further the multidimensionality of local crimes.

Blood, Diamonds, and Fish
The Gulf is rife with forgeries and contraband smuggling. In Lagos, one survey of pharmacies recounted by Naím found that 80 percent of merchandise was counterfeit. The UNODC estimates the annual value of fake and low-quality antimalarial medications at more than $400 million. Cigarette smuggling from the Gulf of Guinea to Europe and North America, in Le Sage's accounting, approaches low-end estimates on the region's cocaine trade. Guns, too, are a problem, circulating with relative impunity. By some estimates, such as that provided by Le Sage, Africa houses 20 percent (100 million) of the world's stock of illegal small arms and light weapons (SALW).[60] Grant writes that the trade in SALW is particularly worrisome in West Africa, where she estimates 7 million firearms are in circulation.[61] Naím argues that the region has also emerged as a producer of craft firearms. Ghana, for example, boasts more than two thousand manufacturers selling pistols for as little as six dollars.[62] The smuggling and production of SALW, as from Angola into Nigeria, makes regional insurgencies "almost intractable," in Onuoha's estimation.[63] Liberia funneled weapons to the Revolutionary United Front in Sierra Leone, as did Burkina Faso and the Ivory Coast. Naím adds that Togo was likewise a staging ground for SALW transshipping to the National Union for the Total Independence of Angola.[64] Blood diamonds, themselves a smuggled commodity, have been used routinely to finance regional weapons procurement.[65] The diamond trade from Sierra Leone has even been alleged of enriching Al-Qaeda, though the evidence has been disputed.[66] Grant, for one, relays an estimate that Al-Qaeda earned $15 million in the trade.[67] Doug Farah, reporting for the *Washington Post*, records that the organization, after its assets were frozen by the United States in 1999, leveraged connections in Liberia to shelter millions of dollars in diamonds in Sierra Leone.[68] The diamond trade, bound up as it is in gunrunning, commodity smuggling, and money laundering, is representative of yet another multidimensional transnational trade. Consequently, smuggling is, in Le Sage's words, "deeply tied to African civil wars" and insecurity. Even more so, he continues, the trade has turned parts of the region into "a 'duty free' port for organized crime."[69]

Despite the obvious appeal of diamonds, there is a less flashy but equally (if not more) important illicit trade taking place in West Africa: fish.

In coastal communities, human security includes the ability to fish for sustenance and profit. Fish account for more than half of all protein consumed by those in the Gulf of Guinea, the poorest 40 percent of whom depend critically on this supply.[70] Illegal, unreported, and unregulated (IUU) fishing challenges this dependence. A report by the International Labour Organization (ILO) concluded that West Africa is "particularly vulnerable to illegal fishing," and that more than a third of catches were illegal in the time studied.[71] Rear Admiral Osinowo of the Nigerian Navy regards these waters as the most abused on the planet.[72] The European Union reinforces such a claim, relaying in their fact sheet a World Bank estimate assessing the annual damage to coastal economies at $350 million.[73]

As we should by now expect, ships engaging in IUU fishing off West Africa have also been documented as perpetrators of other crimes, including forced labor. European and Asian trawlers in West African waters have become increasingly predominant, and trafficking victims have been detected on lakes in Ghana.[74] Meanwhile, Osinowo notes that IUU fishing and related toxic dumping pose serious risks to the socioeconomic health of coastal communities, and environmental degradation is a push factor for migration.[75] More than 1.5 million tons of petroleum has spilled in the Niger Delta, as noted by Lisa Otto of the University of Johannesburg, making it one of the most oil-drenched places on earth (and an obvious correlate to the physical broken windows of chapter 2).[76] Moreover, as noted in commentaries on Somali piracy, the loss of traditional fishing grounds may be connected with an increase in the likelihood of piracy.[77] And while some IUU fishing is the product of local artisanal fishermen, organized crime is also a contributor. In recognition of IUU fishing's relationship to crime and piracy, Chatham House notes that it may even be "far more important in West Africa than piracy."[78]

Finally, money laundering, as we saw in the Caribbean case, is an inevitable compatriot of these illegal trades. Nigeria, in part for its role as a gateway for foreign investment in West Africa, is listed by the United States as a country of primary concern for money laundering. Angola, Benin, Côte d'Ivoire, Ghana, and Sierra Leone are also noted of concern, while Cameroon, Republic of Congo, Equatorial Guinea, Gabon, Liberia, São Tomé and Príncipe, and Togo are among those monitored by the U.S. State Department. Elements of West African money laundering are familiar to our Caribbean case, including bulk cash smuggling, some of which likely takes place using maritime routes given the proliferation of maritime illicit conveyances. Money laundering is also bound up in the smuggling of commodities from West Africa—I have already noted the trade in blood diamonds and fish. Ivory, too, is a major trafficked commodity through Cameroon, Nigeria, Angola, and Côte d'Ivoire.[79] Like narcotics and blood diamonds, ivory and other poached products are occasionally used as cash surrogates to purchase weapons and other illicit goods and services. The open borders and trafficking infrastructure that supports these and other laundering transactions are open to exploitation by all transnational criminals, including terrorist organizations, which have repeatedly used Africa as a "safe haven."[80]

Oil and Political Violence
One of the most predominant crimes in the Niger Delta is illegal oil bunkering, or theft from pipelines. The rise of that crime was in part made possible, as the piracy scholar Martin Murphy notes, by the proliferation of small arms in the Warri region.[81] An Agence France-Presse report detailed daily losses of 150,000 barrels to the theft, representing 7 percent of Nigerian production per an estimate relayed by Otto. Her analysis of media reports estimated annual losses in Nigeria at $12 billion, or 20 percent of GDP.[82] One of the most interesting lineages of this oil theft is its connection to political violence. According to Patrick Dele Cole, former Nigerian ambassador, illegal bunkering began "because of the political need to raise a lot of money during the elections," an idea that was "sponsored and maintained by our political leaders."[83] As Vreÿ notes, oil theft is bound up in an intense competition for resources in which political actors compete for access against other criminal organizations. Control of territory, whether by criminal syndicates or militias, is required to tap and guard wellheads. If local neighborhoods and police cannot be avoided, they may be bribed in much the way narcotics traffickers co-opt communities.[84]

The Gulf of Guinea acts as the downstream distribution hub for this petroleum, where tankers offshore receive deliveries of oil from smaller barges. Here, too, the political-criminal dynamic is on display. One pirate tanker, the *African Pride*, intercepted by the Nigerian Navy, was allowed to offload more than 10,000 barrels of crude to a different pirate tanker while under guard. Two rear admirals eventually lost their jobs in the scandal, but the incident highlights the potential for the complicity of political (government, naval, and business) actors in the crime.[85]

Self-defense groups, militias, and terrorist organizations, as chronicled in chapter 1, often rise to fill the voids left by government inaction or malfeasance and are directly fueled by petroleum. The Movement for the Emancipation of the Niger Delta (MEND) is one such example. It is an outgrowth of political grievances, notably the inadequate distribution of revenue from locally drilled oil and the resulting pollution. Per numbers from Murphy, 40 percent of Nigeria's GDP, more than 80 percent of federal revenues, and virtually all export earnings come from such oil while little investment makes its way back.[86] The group is also criminal, though, perpetrating crimes with the veneer of a political agenda. MEND has since earned the distinction, in Otto's estimation, of being among the leading insurgent group active in criminality at sea.[87] That MEND uses communal issues as a basis for the "intense criminality [that] dominates its practical existence and activities" is tribute to the multidimensionality of insecurity and the importance of communities in framing security issues.[88]

MEND has targeted the petroleum industry as a symbol of federal overreach and has kidnapped workers and tapped and destroyed pipelines.[89] This also includes attacks on offshore oil platforms such as one in 2008 on the *Bonga*, a floating production, storage, and off-loading unit. That incident, according to Kamal-Deen, helped force an agreement between MEND and the government in 2009.[90] That MEND's leadership benefited from this amnesty disproportionately to the rank-and-file, that the organization's leaders have disputed influence, and that MEND's activities were often capital-driven in nature have all complicated that agreement. MEND's structure is characterized by a familiar amorphousness, and Otto writes that that the group is likely "highly fractious," with multiple franchises operating independently under the MEND brand.[91] Murphy writes that MEND was not a corporate, formally structured group but rather a "loose coalition of armed militias," some driven by very local considerations and "grievances."[92] What leadership did exist interwove both criminal and insurgent components, so much so that it wasn't always easy to differentiate the two camps. And as Le Sage notes, plunder can slowly displace politics as an operating principle. As such, he continues, similar to what we explored previously regarding the FARC and *bacrims*, even in instances of political settlement, "insecurity and criminality tend to linger."[93]

Vigilantism thrives throughout the region.[94] In Onitsha, in southeastern Nigeria, a businessman refers to one such group's (the Bakassi Boys) relentless pursuit of armed robbers as "the only thing that makes us sleep with full eyes closed."[95] Perception plays a vital role in the creation and support of such organizations. When communities become so wary of authorities that they avoid entirely interacting with law enforcement, they are forced to look for stability in what Le Sage calls "nonstatutory arrangements." Nothing could more directly speak to the consequences of losing a community's trust, nor could there be greater evidence of the value of community-based police solutions. And while efforts to modernize Africa's militaries have increased professionalism, local police remain plagued by accusations of rampant corruption and violations of human rights.[96] It was in part this "popular perception that the federal state [had] failed as a security guarantor" that made the emergence of the Bakassi Boys possible.[97] Criminal gangs across Nigeria's Rivers State, often sponsored by local politicians, operate with "near-total impunity," in Murphy's words. Such politically sanctioned gangs helped spawn the earliest leaders of MEND, as well as the Niger Delta People's Volunteer Force and the Niger Delta Vigilante.[98] As we have seen, the belief that law enforcement is ineffectual can be enough to threaten such fundamentals as the state's monopoly on the legitimate use of force.

Irregular Migration
Human trafficking and smuggling are bound up, as we saw in chapter 3, in a wider transnational web of actors, routes, and crimes. And like much illicit activity in the Gulf of Guinea, it is no surprise that irregular migration in West Africa predominates on the sea. Human trafficking in West Africa has worsened the criminality that breeds maritime insecurity and, as stated by Chatham House, is "closely related to piracy and robbery" given the shared reliance on weapons trafficking and money laundering.[99] In this regard, it is truly multidimensional, and only by placing irregular migration in a broader context can we hope to understand its role in maritime insecurity.

The push and pull factors leading to irregular migration, and the risk factors of human trafficking and exploitation (all explored in chapter 4), find echoes in the Gulf of Guinea. As in the Caribbean, a great deal of trafficking is intraregional, and most West African countries report that trafficking victims were found subregionally according to the United Nations. Major flows, per Grant, include "from Benin, Burkina Faso, Ghana, Mali, Nigeria, and Togo to the Congo, Côte d'Ivoire, Equatorial Guinea, and Gabon. In Togo, major flows are from Benin, Ghana, and Nigeria."[100] Le Sage identifies major human trafficking and smuggling flows emanating from Nigeria, Ghana and Benin in particular.[101] Globally, West African trafficking victims are transported on international flows and compose a sizeable portion of those detected in Europe. In West and Central Europe, 14 percent of detected victims (around three thousand individuals) came from West Africa, according to one UN report, and West African trafficking victims were found in at least twenty countries across the continent. Only around 1 percent of all victims reported to the United Nations in West and Central Europe originated from elsewhere in sub-Saharan Africa. Nigerians are particularly victimized. They were detected in more than a dozen West and Central European countries, where the United Nations estimates they make up a little more than 10 percent of victims.[102] Individuals from Cameroon, Ghana, Guinea, and Sierra Leone have also been detected throughout Europe. The UNODC estimated in 2007 that revenue from transfer fees for smuggling from West Africa to Europe totals $300 million annually—and that figure has only likely increased. The flow also operates in reverse, with victims trafficked into West Africa. Filipino victims have been reported in Côte d'Ivoire, and Chinese victims have been detected in Ghana.[103]

The ILO calculates the number of forced laborers in Africa at 3.7 million (18 percent of the global total) and the annual profit from this labor at just in excess of $13 billion (higher than in Latin America and the Caribbean). Africa trails only Central and Southeastern Europe in the prevalence of exploited laborers, according to the ILO.[104] The United Nations similarly notes that Africa recorded a higher rate of victims trafficked for forced labor than other regions.[105] This is true of children as well. In Côte d'Ivoire, the ILO estimates that more than 25 percent of children participate in the economy, leaving them vulnerable to exploitation. Lake Volta, in Ghana, is a magnet for children exploited in the fishing industry.[106] Trafficking victims are also disproportionately children. Africa's share of child trafficking is greater than the world average, according to UN statistics; more than 60 percent of detected victims in the region are children. In Côte d'Ivoire, the United Nations noted in one report, the only cases of trafficking they detected pertained to children. In Nigeria, minors account for half of victims, while Ghana and Sierra Leone report that a majority of trafficking victims are children.[107]

Of the estimated 3.7 million African victims of exploitation, approximately 800,000 are engaged in forced sexual exploitation, 2.5 million in forced labor, and 400,000 in state-imposed forced labor.[108] The United Nations estimates that just fewer than 50 percent of victims in Africa and the Middle East were trafficked for forced labor and more than one-third for sexual exploitation. The remaining victims were trafficked for crimes such as use as child soldiers or in ritual practice.[109] In all categories, that is nearly double the figures attributed to the Caribbean and Latin America by the ILO. The ILO also estimates annual profits for sexual exploitation, which in Africa they place at approximately $8.9 billion. For domestic work, they estimate profits of $300 million, and for nondomestic work $3.9 billion, totaling $13.1 billion.[110]

I note extensively in part 2 the community and public health ramifications of irregular migration streams. This was on acute display in August

2014, when the Liberian Coast Guard helped enforce an unpopular quarantine of Liberia's West Point neighborhood. The region, "a sprawling slum with tens of thousands of people," much of which is "made up of shanties separated by alleys wide enough for only one person," was cordoned off in an effort to stop the spread of Ebola.[111] The move, a manifestation of the global fears of infectious contagions, was also representative of the need to address issues of human movement in partnership with communities, not in opposition to them. And while Ebola dominated headlines in the summer of 2014, parts of Africa continue to be ravaged by a range of infectious diseases. HIV/AIDS, for example, remains a leading cause of death on the continent, and human trafficking has been linked to its spread. I also note, in this chapter and in chapter 4, the impact of environmental degradation on irregular migration. In West Africa, illegal logging, appropriation of forestland, and unregulated charcoal making are all present threats to local ecologies that precipitate migration.[112]

As with narcotics trafficking, human trafficking is not West Africa's most visible security threat. Of the top twenty migration corridors identified by the International Organization for Migration, only one African route makes the list, from Burkina Faso seaward to Côte d'Ivoire.[113] On the U.S. State Department's ranked index for trafficking in persons, most Gulf of Guinea countries are listed as Tier Two, with some occasionally hitting Tier Two Watch List (like Angola in 2014 or Liberia in 2017) and Tier Three noncompliant status (like Equatorial Guinea). Nevertheless, as I underscore in part 2, crimes, criminals, and the vectors through which they operate are interconnected. Until a theoretical framework is presented that coherently places irregular migration within the wider context of West African littoral insecurity, combating the trade in people will remain ad hoc and underemphasized.

Corruption and State Weakness

Corruption in West Africa is widely documented. Its most relevant element for this study, however, is the social psychological component, as seen in chapter 5. In an article on piracy, Maj. Eero Tepp of the Estonian Army writes that a "permissive political environment" is one of the noteworthy consequences of corruption.[114] This permissiveness creates an atmosphere conducive to disorder whereby corrupt officials lend protection to allied criminal elements. The result, as in the Caribbean, is the questionable legitimacy of political or security establishments. For example, despite the overt criminality of Niger Delta militants, Murphy considers it undeniable that the perception of the oil industry as a tool propping up corrupt politicians helped groups like MEND substantiate their actions. He further notes that narrowly defining the issues we've been discussing as exclusively criminal, and thus responding with exclusively law enforcement approaches, is problematic "in the absence of a shared appreciation of *what criminal means*," which can breed further disillusionment.[115] Such problems cannot be tackled without a shift in how security apparatuses conceptualize criminality in the littorals.

Corruption is also related to state weakness, which has contributed to the growth of crime in the Gulf of Guinea. Many states in the region, as the UNODC highlights, score in the bottom 10 percent of both rule of law and political stability rankings.[116] Even in Nigeria, West Africa's largest country by demography and economy, Tepp writes that sustained funding shortfalls for all agencies involved in maritime enforcement (the navy, coast guard, and police) has resulted in readiness crises, ineffectual constabulary enforcement, and a "vast network of river transportation routes" that is hardly policed.[117] Although the largest naval force in the region, at around 18,000 personnel, the Nigerian Navy's capacity is so degraded that Tepp considered it doubtful any vessel larger than a small patrol boat was serviceable in 2012. Subsequent U.S. Coast Guard cutter transfers (noted above) have no doubt improved the situation, to a degree. Still, as judged by the ICG, pirates and criminals are often more seaworthy than the navies set to combat them. That ineffectual security sectors also operate in a "climate of impunity," employing torture and intimidation, further weakens their effectiveness while generating an attitude of fear and resentment toward them.[118] As we saw with narcotics, the resulting political space afforded by weak or permissive states acts as a surrogate for the "physical sanctuary of coves, creeks, and swamps" where malicious maritime actors have historically settled.[119]

Finally, as with vigilantism, state weakness is also manifested in the growth of private maritime security companies (PMSCs). Some companies have begun employing armed guards to shepherd ships through dangerous waters, though this has proven a controversial move, and the record of embarked security is mixed.[120] Some fear it will result in escalation with pirates and criminals, and others note a number of legal questions regarding the presence of armed personnel onboard a ship flagged to one state in the territorial waters of another.[121] Yet state weakness and corruption has often meant that distress calls from ship operators in West Africa go unanswered, by the ICG's count, as much as "eight times out of ten."[122] It is then no surprise that the continued use of PMSCs remains likely, especially because the presence of patrol boats (even non-navy contractor vessels) seems to aid in deterrence.[123] After oil, private security is among the largest sectors in the Nigerian economy, according to Vreÿ.[124] As Major Tepp notes, "private navies are on the rise," and their relation to and influence over weak states is still a matter of debate.[125] While the presence of PMSCs may signal a growing willingness for added security among those able to pay, at a broader level the predominance of PMSCs may only reinforce the perception that states in West Africa do not retain the capacity to project power in their littorals.

Snapshot
The Gulf of Guinea exhibits the same basic traits—multidimensional threats and the relevance of context—that make the Caribbean so amenable to an application of the broken windows theory. Speaking to multidimensionality, Le Sage writes that the myriad irregular challenges Africa faces "are not isolated phenomena." Instead, he notes, they act on one another and create a "vicious circle of insecurity."[126] We know from part 2 that trafficking infrastructure is easily co-opted for a wide range of needs or goods. This is likewise the case in the Gulf of Guinea, where crime, insurgency, and corruption serve as "ready networks" for use by criminals, terrorists, insurgents, or others in pursuit of "weapons proliferation, illicit finance, and illegal movement of men and material."[127]

Speaking to context, we have seen that local violence, fed by transnational flows, permeates the environment in which so many live. Many in West Africa, Le Sage writes, are daily subject to "all forms of criminal violence ... including robberies, assaults, carjackings, kidnappings, and sexual violence, with little recourse."[128] Kamal-Deen alludes frequently to this context. Poor health care, difficult living conditions, and schools in disrepair are literal broken windows, precipitating "disillusionment and frustration," in his words, which we know from chapter 2 to be corrosive for community resilience. In such an environment, "resentment would be at its height," to quote Kamal-Deen, resentment which could precipitate disorder, criminality, and armed conflict. All of this is familiar from both my discussion of the broken windows theory and its insight for the Caribbean, suggesting that the precepts of broken windows are just as applicable in the Gulf of Guinea as in the Western Hemisphere. Yet this particular "malignant environment" has produced one feature unlike what we saw in the modern-day Caribbean—rampant piracy.[129] Demonstrating how this issue can be integrated into a broken windows framework is critical for determining the theory's generalizability.

Piracy

Piracy was likely the first act criminalized under international law—*hostis humani generis*, enemies of all mankind—and, as Otto narrates, its cross-jurisdictional nature helped develop the very notion of the high seas as the common heritage of humanity.[130] Yet piracy today in the Gulf of Guinea is not that of Hollywood lore, and contemporary definitions prove insufficient in addressing it. The most widely used definition of piracy is set out in Article 101 of the UN Convention on the Law of the Sea (UNCLOS), the international benchmark for maritime law. According to the convention, piracy is an action that can only take place on the high seas, beyond the territorial sea maximum of 12 nautical miles (nm) from the baseline. The high seas include the contiguous zone (the next 12 nm), the exclusive economic zones (up to 200 nm from the baseline), and the remainder of international waters beyond. Piratical acts that take place within the territorial sea or internal waters are de jure not piracy but merely armed robbery at sea.[131] Otto notes that, while UNCLOS's intent was to preserve coastal state sovereignty, the resulting arbitrary distinction between piracy and robbery where such actions frequently take place between the territorial and high seas hampers enforcement efforts.[132] In the case of Somali piracy, this debate has been largely sidestepped. Somalia's status as a failed state and the great distance from the coast at which Somali pirates operate negate many of these legal issues (while opening many others, such as where to try pirates when they come from a failed country). In the Gulf of Guinea, however, armed robbery is prevalent in territorial waters—pirates in Nigeria operate from a web of thousands of inland creeks and streams (more than three thousand, by Tepp's count[133])—so the dogmatic delineation in legal language has become a pronounced hindrance.

A range of definitions has been proposed to correct the artificial distinction when acts of violence occur along the divide of the territorial sea. The International Maritime Bureau (IMB) emphasizes intent over geography, folding robbery at sea and piracy together.[134] The IMB's definition, favored by Onuoha over that of UNCLOS's, defines piracy as "an act of boarding or attempting to board any ship with the apparent intent to commit theft or any other crime and with the apparent intent or capability to use force in the furtherance of that act."[135] This reflects the wider Gulf of Guinea scholarship, which avoids geographically conditional definitions. Even the UN Security Council's choice of rhetoric includes robbery at sea as de facto piracy.[136] While this may help ensure that counterpiracy efforts avoid UNCLOS's self-imposed strictures, Otto notes that existing legal constructs remain significant institutional barriers. Otto concludes that a more dynamic construction is imperative in order for piracy to be considered "holistically rather than in individual parts specific to the distance from the coast at which the crime takes place."[137] Here I adhere to the above functional definitions, recognizing that for all practical accounts, the threats, actors, and motivations behind piracy and robbery at sea are, as Kamal-Deen summarizes, "largely the same despite the legal distinction."[138]

Formulating a more realistic definition of piracy cannot be divorced from the act's inherent multidimensionality, including the often-downplayed nexus between piracy and political violence. In 2008 and 2009, as Murphy notes, militants from the Niger Delta mounted cross-border attacks into the Bakassi peninsula (governed by Cameroon) as well as Malabo (the capital of Equatorial Guinea).[139] Political unrest in the Bakassi peninsula in particular may be an influence, according to Vreÿ, while Kamal-Deen notes that the Bakassi Freedom Fighters may be linked to some attacks.[140] Of Nigerian pirates, Dirk Steffen, director of maritime security at Risk Intelligence, notes that they are "the same youths who face both oppression and extortion by corrupt security forces in the Niger Delta."[141] The interconnectedness of piracy, criminality, and politics leads Vreÿ to identify a "pirate-insurgent-criminality threat." He notes that piracy in the Gulf of Guinea is often a display of a "political-criminal link," just as with oil bunking ashore.[142] And just as political unrest serves as justification for theft on land, so too do pirates masquerade as insurgents to justify their pillaging.

Perhaps the surest sign of the relationship between piracy and politics is the hypothesis that pirate attacks increase around elections. The managing director of Risk Intelligence, the director of Chatham House's Africa program, and the IMB have all advanced this theory. This trend appears to be supported by piracy patterns (though 2011 is an outlier), including an increase in attacks leading up to the 2015 elections. This brings important questions to the fore about piracy's role as a "campaign finance mechanism," which remains an area of open research.[143] As Tepp notes, there has always been a relationship between piracy and local political conditions, though he stops short of addressing piracy as overtly political.[144] Otto goes still further, questioning (rightly, in my opinion) whether UNCLOS's apolitical categorization of piracy accurately captures the varied motivations of Gulf of Guinea actors.[145]

This primer on the relationship between piracy and politics bears directly on the hybrid nature of crimes in the littorals. The difficulty in constructing a broad theory of maritime security—one that is not just counter-something—is facilitated by the self-imposed strictures of dividing groups into political or criminal, which is particularly pronounced in the case of piracy. Meanwhile, police infrequently have the capabilities necessary to combat threats in the maritime sphere, particularly those with criminal patinas but the capacity (and perhaps the will) to exercise political violence. Still, leveraging military tactics or matériel is not an easy solution. As I explored with the conversation on urbicide, militaries tend to ignore the communal and contextual natures of security, resulting sometimes in counterproductive methods. This is reinforced in chapter 5 with the conversation on fortress Caribbean and the standoff in Kingston, Jamaica, to arrest Christopher Coke. The solution to this mismatch between police know-how and military capability was melding the precepts of community policing with the might often imprecisely leveraged by navies or coast guards. In the following, by identifying the multidimensional and contextual elements of Gulf of Guinea piracy, we might discover that piracy is similarly amenable to framing in a broken windows context.

The Numbers
As Murphy notes, Nigeria had a piracy problem a decade before widespread Somali piracy burst to the fore. And Gulf of Guinea piracy appears to be enduring as well, in a way Somali piracy may not be. In 2012 it again overtook Somalia in number of attacks (due to a combination of mildly rising West African numbers and rapidly shrinking figures off the Horn of Africa). According to figures from Kamal-Deen, attacks nearly tripled between 2005 and 2007, and swelled yet again between 2010 and 2013 after a short decline (2008–9).[146] In 2016, Gulf of Guinea piracy reached a high-water mark as a percentage of attacks globally (about 30 percent).[147] During the heyday of the Niger Delta insurgency, until the 2009 settlement, ICG analysis notes that incidents were often worst off of Warri, Port Harcourt, and Calabar.[148] The spike in pirate activity following the lull precipitated by the 2009 accord between Nigeria and MEND demonstrates that piracy persists, even in the face of political settlement, as an increasingly popular option for financial gain. In the aftermath of this agreement, insurgents have not so much joined forces with pirates as simply become pirates themselves, in Kamal-Deen's estimation.[149] Recall, in chapter 4 I detailed the relationship between terrorism and transnational crime—particularly the potential for the former to migrate to the latter. I noted that the skillset required to operate as a terrorist is often exportable to the less ideologically motivated elements of international crime, creating linkages and potential nodes between terror and transnational crime.

Moreover, as Le Sage notes, piracy may be "contagious," producing a copycat effect elsewhere.[150] While the progression from pirates' political motivations to financial motivations (above) represents a multidimensional element of West African piracy, this contagion serves as a fundamental example of the impact of context. Near Lagos, for example, the ICG reports that piracy increased commensurate with spikes in the delta, with attacks taking place well within range of shore. Acts of piracy also migrated upstream in the delta, they noted, where waterways such as the Port Harcourt–Nembe axis and the Calabar River became hotspots for robbery.[151] Today Nigeria remains the epicenter of West African piracy. In 2016, according to the IMB's Piracy Reporting Centre, of the approximately 191 incidents they noted worldwide, 36 took place off Nigeria alone.[152] Nigerian waters, by Kamal-Deen's account, have seen the greatest number of incidents directed against offshore oil rigs anywhere in the world.[153] This has profound consequences for the cost of shipping in the region as well as the dollars lost to regional countries (estimates relayed by Osinowo are up to $2 billion annually[154]). Further, numbers for West African piracy are likely underreported to a greater degree than elsewhere (Kamal-Deen guesses by as much as half[155]), giving greater significance to these figures. That piracy breeds more piracy in part merely by inspiring copycats is indicative of the crime's viral quality and is emblematic of the epidemiological representation of crime that broken windows produces, down to the theory's adoption in public health. This contagious feedback loop is likewise expressed in Philip Gosse's piracy cycle, which outlines a familiar evolution of maritime piracy, from small scale, opportunistic attacks to larger and more organized operations.[156]

Piracy's contagion is on display across the Gulf of Guinea. While piracy emanates primarily from Nigeria, it is quickly becoming a regional issue. This is difficult to measure in part because states likely underreport acts of piracy. Nevertheless, we can at least discern trends. Attacks on oil tankers, for example, have been registered throughout the "Ghana-to-Angola and Nigeria-to-Côte d'Ivoire axes."[157] In the summer of 2013 a Maltese ship was hijacked off of Gabon, the first such incident in its waters. Piracy from Guinea appears to have spilled into Sierra Leone, which Kamal-Deen regards as a "portent" of organized crime.[158] Equatorial Guinea reported its first major attack in February 2009, according to ICG, which was particularly telling. In what was first thought to be an attempted coup, they recount, approximately fifty armed and poorly informed assailants launched a raid on the presidential palace in Malabo. The assailants later admitted they thought the property was a bank or was in possession of large sums of money.[159] Some of this piracy outside of Nigerian waters is born of the balloon effect. As Nigeria made gains countering piracy in the late 2000s, Benin went from one reported attack in 2009 and zero in 2010 to more than twenty in 2011 before returning to few or no attacks in most years, according to the IMB.[160] Greater enforcement off of Benin (with Nigerian help) was then followed by a rise in attacks in Togo (and Nigeria once again), according to statistics recounted by Osinowo and Kamal-Deen.[161]

Piracy is not evenly distributed in the Gulf. Between 2012 and 2014 Togolese and Nigerian piracy accounted for three-quarters of incidents in West Africa (the bulk being Nigerian, with a large spike for Togo in 2012 specifically). Côte d'Ivoire (and, to a lesser extent, Sierra Leone) have become of increasing concern, and Guinea has established itself as a secondary enclave of piracy, by Kamal-Deen's assessment. In 2016 Nigeria alone represented around 65 percent of Gulf of Guinea attacks, according to data from the IMB. Alternatively, Angola only reported its first major hijacking in February 2014, according to Kamal-Deen, although the country has seen periodic attacks (a peak, so far, of four in 2006 according to IMB data).[162] Piracy off of Cameroon has meanwhile declined since its peak in 2009–10. Despite local shifts, however, overall Gulf of Guinea pirate reports have proven remarkably durable over the past twenty years. Steffen relays that the Gulf of Guinea has averaged, in recent years, between eighty and one hundred attacks annually.[163] Even including the obstacles of data collection, transparency, and figures lost in the definitional dispute between robbery and piracy, the trends are clear: the Gulf of Guinea is one of the most insecure waterways in the world.

The Tactics

The word "piracy" often evokes images of Johnny Depp (or Errol Flynn) swashbuckling in Tortuga, or Capt. Richard Phillips of the MV *Maersk Alabama*, held hostage for days. In West Africa, however, piracy takes a much different form. The ICG identifies three types of maritime criminality there: "the spread of political gangster-ism from the Niger Delta to the Bakassi peninsula; seaborne raids; and increasingly sophisticated acts of piracy."[164] The latter two can also be described as opportunistic vice organized piracy. According to Tepp, opportunistic piracy is "not as much a way of life as a crime of opportunity," lending the issue particularly well toward an evaluation through a broken windows perspective.[165] If crime is opportunistic, that means it is rooted fundamentally in the perceptions of those who perpetrate the acts. And, as I detail in chapter 2, if perceptions can be altered, so can actions. Moreover, Kamal-Deen categorizes most Gulf of Guinea piracy leading up to 2005 as opportunistic, when robberies were mostly "subsidiary activities." He notes that this label is not a commentary on capabilities but rather about the centrality of piracy for an organization.[166] Such robberies, stealing easily portable cargo and equipment, can net between $10,000 and $15,000 in one attack, grossing as much as $1.3 million annually, according to reporting from the *Washington Post*. Coordinated attacks, such as those choreographed against oil tankers, can net nearly triple that figure according to that same reporting.[167] In such attacks as many as thirty men in up to five speedboats converge on a target at speeds of 60 knots according to ICG reporting.[168] (Speedboats, which can often outrun naval craft, were already in use by littoral insurgents and were easily repurposed for piracy.[169]) This represents a surgical strike requiring operational intelligence, as described by Kamal-Deen, unlike the prowling technique often employed in Somalia. Groups including MEND, the Niger Delta People's Volunteer Force, and the Niger Delta Vigilante have been identified with this type of attack.[170] Oil theft, as noted by Osinowo and Otto (among others), also includes the involvement of organizations from as far as Eastern Europe or Asia.[171] This connection to transnational crime can be seen in the hijacking of the *Duzgit Venture*, as recounted by Kamal-Deen, during which pirates acted "in cahoots" with another band of actors several thousand kilometers away. In an attempt to meet their coconspirators, Kamal-Deen notes, the hijackers prodigiously navigated their prize across the territorial waters of five other nations.[172] Yet, as I detail in chapter 4, even organized crime fits into the broken windows' theoretical lens. At the most granular level, transnational crime does not exist in a vacuum—all crime happens in someone's community.

Murphy traces the origins of organized operations to the 1970s, as the oil industry took hold, bringing with it an upsurge in shipping for construction materials (the "cement armada").[173] Then, piracy was centered on Lagos. As the petroleum industry migrated southeast, stirring feelings of disenfranchisement, piracy followed suit. This antecedent to contemporary piracy exhibited several characteristics that have proven durable. Robbery, for one, was no longer the only apparent motive. Sabotage (and, to a lesser extent, kidnap and ransom, as Otto points out[174]) became prominent, and it had become evident that piracy was increasingly part of a complex set of criminal, political, and tribal concerns. Violence also became a rising motif at this stage, according to Murphy's recounting. Nigerian pirates are alleged to be more violent than their Somali counterparts, usually for tactical reasons (see, for example, Osinowo, Murphy, and

Kamal-Deen). The former aim to facilitate an efficient robbery, while the latter employ kidnap for ransom and thus need live victims.[175] The use of violence to ensure complicity may also be a result of a lack of safe havens from which Nigerian pirates can conduct ransom negotiations, as posited by Kamal-Deen (among others).[176] This violence is consequential because successful boardings are more common in the Gulf of Guinea than elsewhere in Africa, by Gilpin's account, and the region is broadly noteworthy for the high rate of successful attacks and shipboard kidnappings.[177] The use of automatic rifles and rocket-propelled grenades, Kamal-Deen relays, is not uncommon in such attacks.[178]

As our understanding of broken windows would suggest, these attacks have a reverberating effect on local communities. In particular, piracy has impacted West African fishing more than any other local sector, according to Tepp. In a six-year period, Nigeria documented nearly three hundred pirate attacks on fishing boats. Fishing vessels are also targeted for hijackings. Pirates use stolen trawlers' expanded ranges to operate as mother ships from which to stage attacks on oil tankers farther out to sea as well as for use as a ruse to approach oil tankers.[179] Recall the importance of local fishing to the livelihood of regional communities. As we saw in the discussion in chapter 4 on Jamaican fishermen, communities like Forum, Jamaica, mentioned in chapter 3, are bellwethers of littoral security. It should be no surprise that counterpiracy operations frequently see benefits in the fight against IUU fishing as well, an explicit endorsement of piracy's multidimensional quality.[180] Yet navies need not be only reactive. By engaging proactively with littoral communities, as community policing policies would recommend, authorities are afforded the opportunity to address problems when first detected by communities themselves—before they become regionally entrenched.

Communities also feel the consequences of poor maritime security strategies. Journalist Stephen Starr, writing for West Point's Combating Terrorism Center (inadvertently underscoring the value of a theory that joins criminology and maritime security) expresses that the "line between piracy and conventional criminality is becoming increasingly blurred in Nigeria, resulting in the 'cross-pollination' of the country's naval and police forces."[181] This cross-pollination is a reference not only to the hybridization of security threats but the hybridization of responses to those threats, which, as we have seen in chapters 1, 4, and 5, can be nearly as corrosive for community safety than the threats they combat.

Given piracy's linkages to political violence, organized crime, and human security, it is not inconceivable that piracy and terrorism may become linked. Pirates maintain capabilities that many terrorists would envy. Although such a relationship remains a matter of speculation,[182] an attack on a pipeline in February 2012, recorded by Kamal-Deen, sparked concerns that Boko Haram might begin to target petroleum infrastructure, including maritime targets.[183] Similar concerns, with some more supporting evidence in chapter 4, suggest that such crime–terror partnerships are possible. Fluid criminal dynamics and the potential for transnational criminal organizations to co-opt existing infrastructure make the terror–piracy nexus worth monitoring. Aside from acts of piracy committed in the maritime space, terrorists may also adapt the infrastructure of piracy for use entering and exiting Africa in obscurity; for example, the Caribbean could be used as a staging ground for acts elsewhere. As we saw in the Caribbean, the capabilities inherent in human trafficking, narcotics trafficking, and now piracy overlap. As we also saw in part 2, these links are not simply theoretical. While every link may not be of equal strength or maturity, the multidimensionality of maritime crime is a fundamental component of the littoral environment.

Pirates in the Gulf of Guinea threaten good order at sea not in a vacuum but alongside poachers, narcotics traffickers, insurgents, and other criminals, sometimes with the same groups or individuals fitting multiple labels. Chibuike Rotimi Amaechi, former governor of Nigeria's Rivers State, notes that someone might "one day kidnap an oil worker in order to buy a flashy car; the next day he may join a raid by a militant group and, on the third day, hijack a rig to generate cash for his chief or to get jobs" or services for his community.[184] This fluidity runs counter to the usual counterpiracy, counterinsurgency, counternarcotics divides and demands a strategy equipped to discuss the littoral environment as a larger system. Piracy is not an isolated phenomenon; it is one node in a dynamic web of interrelated issues.

BROKEN WINDOWS

In part 2, I detailed that insecurity in Caribbean littorals, seen through two of broken windows' central tenets (multidimensionality and context specificity), evinced the relevance of applying a criminological lens to maritime security. I also identified that the enforcement environment in the Caribbean adds yet more incentive to apply community policing methods in the littorals. Heavy-handed tactics employed by hybridized military and law enforcement units often alienate the neighborhoods they are meant to protect, while the predominance of small actors (each unable to independently address maritime insecurity) necessitates a consensus theory of security to ensure unity of effort. In this enforcement arena, community policing presents itself as a ready solution. The Gulf of Guinea faces similar obstacles, brokering the possibility that it may be amenable to a similar solution.

Enforcement Environment
First, like the Caribbean, West Africa is highly sensitive to perceptions of militarization of the region, particularly with respect to American military involvement. In part for this reason, the United States has not maintained dedicated counterpiracy assets in the Gulf of Guinea, and programs like the Africa Partnership Station (APS) have avoided large footprints. This reluctance toward militarization is also part of the reason for AFRICOM's limited continental footprint (it is based in Stuttgart, Germany). Following the announcement of the creation of AFRICOM, Congressional Research Service Africa expert Lauren Ploch noted that most West African military heads expressed serious reservations about the command. After then-Nigerian president Umaru Yar'Adua hinted optimism about the command in a visit to Washington, Ploch writes, domestic backlash quickly prompted the foreign minister to walk those comments back.[185] Meanwhile, Ghana's president, John Kufuor, expressed relief after a meeting with George W. Bush that the United States was not planning for a robust institutional presence on African soil.[186] There are, of course, other reasons for nations, particularly those with colonial pasts, to fear the presence of foreign military installations on their soil. Yet, just as the fear of urbicide and "fortress Caribbean" plays a role in framing security in the Western Hemisphere, so too does it in West Africa. In Ploch's evaluation, she notes concerns over the potential "militarization of development and diplomacy," with some fearing that the Department of Defense could sideline the State Department and U.S. Agency for International Development on the continent.[187] And this is not solely an issue of American intervention. As in the Jamaican example of Tivoli Gardens, West African states have also "militarized" their responses to criminality. As Chatham House notes, some countries have created special forces units specifically to perform police functions.[188] All of this speaks to the pervading role of context and perception in informing how security solutions are framed, even if regional sentiment is a particularly subtle concept to capture.

Second, as in the Caribbean, there is a proliferation of interested regional actors with competing incentives, complicating the development of a shared framing of security issues. West African organizations have created a veritable alphabet soup of counterpiracy players—ECCAS, ECOWAS, MOWCA, GGC (Gulf of Guinea Commission) and more.[189] In 2008 the ECCAS produced its Integrated Strategy for Maritime Security, which aimed to construct a "common regional framework" for action.[190] The Yaounde Code of Conduct, signed in June 2013, represents an effort to coordinate counterpiracy efforts by dividing the Gulf into alphabetized areas of responsibility.[191] MOWCA, in partnership with the International Maritime Organization, has sought to facilitate an integrated coastal force although progress has been sluggish.[192] MOWCA's membership—it is represented through transport ministers who wield limited influence—and the bureaucratic competition between coast guards and navies have contributed to this lethargy, in the ICG's estimation.[193] A similar effort put forth by the Gulf of Guinea Commission—the Gulf of Guinea Guard Force—also stalled.[194] One of the most successful examples of cooperation came from a limited duration joint cross-border patrol undertaken by Nigeria and Benin in late 2011.[195]

Piracy in the Gulf of Guinea also attracts the attention of international actors. UNODC and the International Maritime Organization have both facilitated regional maritime security training, and the UN Security Council has passed resolutions calling for greater focus on Gulf of Guinea piracy (UN Security Council Resolutions 2018 and 2039). In January 2013, the European Union introduced its Critical Maritime Routes in the Gulf of Guinea program.[196] In the spring of 2014 the Chinese People's Liberation Army Navy visited Cameroon and allegedly participated in counterpiracy drills.[197] Reports in 2016 suggest that China is also providing shore-based counter piracy support to West African states.[198] France has bolstered partnerships with Cameroon, Gabon, Equatorial Guinea, and Congo-Brazzaville, as reported by the ICG.[199] NATO has indicated interest in a greater presence in the Gulf. Even Interpol is involved, providing investigative support, underscoring the hybridized nature of littoral security and the potential role of criminology.

The U.S. Navy has taken steps to offer maritime security assistance for regional needs. In response to the diversity of relevant actors, AFRICOM's naval element (NAVAF) has attempted to "breathe life" into regional agreements and has emphasized joint exercises like Obangame and Saharan Express.[200] To provide aid without sparking sensitivities, the United States also offers support through partnerships and capacity building. The APS and the African Maritime Law Enforcement Partnership (AMLEP) programs are particularly aimed at maritime security assistance. APS, for example, began in 2007 with the deployment of USNS *Swift* (HSV 2) and USS *Fort McHenry* (LSD 43).[201] These ships operate primarily as "floating schoolhouses," as Ploch writes, to partner with local maritime forces.[202] And, given the overlap with constabulary responsibilities, the U.S. Coast Guard is a key partner in APS. AMLEP, meanwhile, is similar to the law enforcement detachments operating in the Caribbean. In this partnership, embarked U.S. Coast Guard teams operate alongside host nation law enforcement, often on American ships, to provide law enforcement training and support.

More broadly, as with Southern Command, AFRICOM's priorities are not generally combat-oriented. Responsibilities in both regions include making gains in what I call the contextual domain, including providing humanitarian assistance and disaster relief and conducting counternarcotics and maritime interdiction operations. Moreover, as introduced in chapter 1, these constabulary operations may require "a major break with conventional doctrinal mentalities," given how they differ in substance and intention from what the Navy typically does.[203]

Gulf of Guinea states have also unilaterally invested in maritime security. Nigeria has installed surveillance posts along its coast in an effort to dissuade piracy while Côte d'Ivoire has stated its intent to expand its navy by forty ships.[204] In June 2014 Equatorial Guinea debuted a "jury-rigged frigate to lead its emerging naval force" while Gabon chose to purchase two offshore patrol vessels during a naval exhibition that year.[205] In 2012 the Ghanaian navy greatly expanded its capacity through the acquisition of two German and four Chinese ships.[206] The rate of growth signals the newfound prominence of regional maritime security and the need for a strategy to guide investment. Yet it is in multilateralism that the need for a coherent strategy is most apparent. Cooperation will be the theme of West African counterpiracy since regional national authorities lack the resources to address the problem and international forces lack the legal authority to act independently. As foreign actors grow ever-invested in counterpiracy, and as regional states invest greater sums in capabilities and organizations, the need to arrive at a consensus strategy of maritime security—one that does not lose sight of the local context and inhabitants—is compelling.

As we can see, West Africa's panoply of security concerns is leading to greater engagement on the part of Europe, NATO, the United States, and China, not to mention all of the regional littoral states. And yet maritime security cannot be achieved simply by detaching more ships to the Gulf of Guinea. As the ICG report argues, the only way to successfully address an issue such as piracy is to see it not as a discrete crime but as a manifestation of transnational organized crime. Consequently, navies in such hot spots are likely to sustain or even grow their constabulary responsibilities in the future. Moreover, as in the Caribbean, West African transnational criminal organizations will continue to diversify into as many product streams as possible. Not only do criminal groups engage in the trafficking of people, narcotics, commodities, and weapons on land, they are also engaged with the pirates plaguing coastal waters. Scholars and practitioners who observe the region agree that, despite the public (and academic) fascination with piracy, "it is but one aspect of a wider problem of maritime disorder," as Murphy summarizes.[207] So, why should counterpiracy efforts exist in a void if piracy does not?

A senior official at Interpol, Yaron Gottlieb, is quoted in Chatham House's report speaking to the "'glocalization' of anti-piracy efforts." Glocalization reflects the intersection of local phenomena with globally sourced solutions. He argues, "As criminals take global ideas and implement them at a local level, addressing the problem requires global solutions that are tailored to local environments and conditions."[208] This is precisely the impetus for the theoretical framework put forth in this book. The decision to investigate a community policing construct is also increasingly relevant. While piracy's impact on global shipping and oil prices may garner the interest of Western powers, maritime insecurity is particularly important for the havoc it wreaks on local communities. In fact, in West Africa, all of the themes of the previous chapters converge—multidimensionality, context, Kilcullen's megatrends, hybrid threats, the need for cooperation, and the shortfalls of evaluating maritime security without reconciling its political and criminal components.

Discourse

Perceptions and attitudes play a crucial role in Gulf of Guinea piracy. As in the Caribbean, a careful review of the literature reveals frequent allusions to the role of context in maritime security. Otto, for example, notes how a "political *culture*" of corruption has precipitated "an *enabling environment* for pirates." Otto also references "the rise of a *culture* of violence," which may inform the character of West African piracy. Otto proposes that the primary difference between Somali and Nigerian piracy is that the former's state weakness turned into state failure, whereas Nigerian weakness has manifested in "a state in which violence is the norm."[209] The ICG, meanwhile, refers to a palpable "criminal culture" in the region. This language of norm setting and perception invites a deeper analysis of the discourse on piracy in West Africa.

Tepp summarizes Murphy's position as an argument that the cultural factors connecting communities to the maritime space is one of several conditions that may make piracy more likely in a given region, including West Africa.[210] While Tepp is skeptical of this reasoning, asserting that "Nigeria does not have a true culture of piracy," he nevertheless agrees that "the crime [now] enjoys social acceptance amongst riverine communities of the Niger Delta."[211] This support comes in part because of the feelings of financial and political marginalization discussed above. Of the seven factors Murphy identifies as contributors to piracy, he includes the role of "conflict and disorder," "permissive political environments," and "cultural acceptability," all of which speak in some form to context and perception, which we know from the broken windows conversation in chapter 2 to play a crucial role in the proliferation of disorder. In Nigeria, Murphy continues, "resentment" plays an instigating role in maritime crimes. Harkening back to our conversation above on the contagion of piracy, Murphy contends it can also turn endemic, with its "corrupting influence" becoming self-perpetuating.[212] Many states even lack "the culture to protect maritime matters," which is "weakly developed" in deference to security threats on land.[213]

Murphy, like many others, argues that we can apply a rational, economic lens to piracy, which he notes is ultimately a business venture. That is a valid lens and has been applied for decades in criminology as well. Yet, as I have illustrated, crime understood purely on the rational actor model fails to emphasize important elements—namely, multidimensionality and environment—that drive behavior in ways rational actor modeling won't capture. As the broken windows theory highlights and as the discourse on West African piracy contagion suggests, disorder breeds further disorder. Failure to engage with communities and address insecurity as it unfolds only increases the cost of intervening over time.

Proposed solutions to piracy in the Gulf of Guinea are equally evocative of broken windows language. The ICG advocates engaging with artisanal fishermen who know their coastal neighbors and could provide intelligence and partnership to maritime authorities (reminiscent of the Colombian villagers in chapter 5, who detected a narco-sub operation and informed local police). The European Union identifies aiding "vulnerable communities" as among its primary values in the construction of a Gulf of Guinea strategy. René-Eric Dagorn of Sciences-Po Paris likewise writes that counterpiracy should involve not only state building but also "society building."[214] Rear Admiral Osinowo even explicitly includes a failure to implement community policing methods as among the major limitations to better addressing maritime insecurity.[215] The sentiment throughout these statements is that crime cannot be disassociated from the contexts in which they take place: their communities.

❖ ❖ ❖

I began this chapter by underscoring the thematic similarities between the Caribbean and the Gulf of Guinea—first in the narcotics industry and then across a range of crimes that affect the lives of many living on the West African coast. I demonstrated that maritime insecurity in the Gulf of Guinea reflects important components of maritime disorder seen in part 2—namely, multidimensionality and the importance of context. Moreover, particularly in the case of narcotics trafficking, many of the same people and networks indeed operate across both regions. In the second half of this chapter I turned to piracy, something we had yet to discuss in serious detail. In this discussion, piracy fits comfortably in the framing provided by the broken windows theory. The theory, underscoring piracy's multidimensional components and relevance to community's and context, yields insights in line with some of the discourse and recommendations in the literature on piracy in West Africa. As a consequence, the broken windows theory provides a useful security lens beyond just the Caribbean Basin. To put a fine point on this assertion, the following chapter explores the theory's use in one final region—Southeast Asia.

The littoral combat ship USS *Coronado* (LCS 4; pictured) sails with the Royal Malaysian Navy frigate KD *Lekiu* (F30) and corvette KD *Lekir* (F26) during Maritime Training Activity Malaysia 2017. *Coronado* is on a rotational deployment in U.S. Seventh Fleet area of responsibility, patrolling the region's littorals and working hull to hull with partner navies to provide Seventh Fleet with the flexible capabilities it needs now and in the future. *U.S. Navy photo by Mass Communication Specialist 2nd Class Kaleb R. Staples/Released*

CHAPTER SEVEN
Evolving Security Conceptions in the Straits

In this concluding case study, let us turn to a conversation on a region that is widely considered to have met its maritime insecurity head on and, along the way, developed some of the most mature literature in maritime security research. By demonstrating the relevance of a broken windows lens even here, we can lend still greater credibility to the argument for the theory's generalizability. As we will see in this chapter, despite dedicated efforts to address and suppress piracy in the Straits of Malacca and Singapore, problems in framing and theorizing regional maritime security have led to obstacles in practice. Moreover, as in the Caribbean and the Gulf of Guinea, we will see in the following that the literature on maritime security in Southeast Asia has slowly, haphazardly, but demonstrably begun to echo sentiments of what a more organized broken windows lens would argue, as theorists continue to wrestle with the durability of piracy. Thus, while chapter 6 provided proof of the theory's generalizability through the discussion of a new issue (piracy), this chapter reinforces the argument for generalizability by analyzing long-standing counterpiracy efforts and demonstrating the value of broken windows theory even in this context.

THE REGION

A strait is a narrow stretch of water connecting two larger bodies—in the case of the Straits of Malacca and Singapore, two very narrow stretches of water, connecting some of the most vital shipping routes in the world. Collectively, these straits constitute the longest and most used waterways in international shipping.[1] The Strait of Malacca runs approximately 600 nautical miles, from the Andaman Sea to the South China Sea. More specifically, the International Hydrographic Organization places the Malacca Strait "between the west coast of Thailand and Malaysia in the northeast, and the coast of Sumatra in the southwest."[2] At its narrowest, the Strait of Malacca is only 1.5 miles wide, situated between Peninsular Malaysia and Sumatra. The Singapore Strait, which bridges the Strait of Malacca with the South China Sea, is even narrower at points, winnowing down to 1.2 miles across, never more than 15 miles wide, as it winds between peninsular Malaysia, Singapore, and Indonesia's Riau Islands.[3] These narrow lanes serve as arteries for the global economy. By tonnage, more than half of the world's shipping traffic passes through the Straits of Malacca, Sunda, and Lombok annually.[4] And according to reports by Japan's Ministry of Land Infrastructure and Transport, regional maritime traffic could soon exceed 100,000 ships.[5]

This traffic carries much more than just goods bound for market in the west. These shipping lanes have deep strategic significance to China, Japan, South Korea, and other rapidly industrializing states in need of Arabian oil and natural gas.[6] And it is not only commercial vessels that rely on the maritime highways of Southeast Asia. American warships likewise employ this causeway to the Middle East, transiting from operations in the Persian Gulf and Arabian Sea to the South China Sea.[7] With the advent of the pivot to Asia and a post-9/11 emphasis on radical Islamic groups, the Straits of Malacca and Singapore have drawn still greater attention from American naval planners. The Bush administration, for example, launched three maritime security initiatives during the 2000s—the Container Security Initiative, the Proliferation Security Initiative, and the Regional Maritime Security Initiative (RMSI)—each with an eye toward Southeast Asia.[8]

As I also note in chapter 1, Southeast Asia's population is heavily clustered in the littorals. Two-thirds of Southeast Asian residents live in or are reliant on the maritime sphere, according to Lt. Cdr. John Bradford, an Olmsted Scholar then studying in Singapore.[9] The waterways that many live and depend on connect three of the largest countries in the world: China, India, and Indonesia.[10] Climate change presents real concerns for such coastally concentrated populations. The combination of coastal congestion and global climate change precipitates real concerns not only of human security but also of regional instability through disruptions to traditional livelihoods and irregular migration. Already Southeast Asia is, according to some, the "most disaster affected area in the world," prompting the establishment of a regional coordination center for disaster relief in Singapore and increasing focus on humanitarian assistance and disaster relief (HA/DR) by local armed forces.[11]

Unlike in the previous case studies, however, the area of the Straits of Malacca and Singapore is dominated geographically by few (only three) littoral states—although other nations along adjacent sea routes and critical user states serve to augment the ranks of interested parties. Yet, even without a wide cadre of directly affected parties, the three littoral states alone—Singapore, Malaysia, and Indonesia—are widely diverse in wealth, size, and governance. Justin Hastings of the Georgia Institute of Technology notes the extent of this diversity: "Singapore, one of the wealthiest countries (with one of the highest levels of state capacity) in the world, is connected to Malaysia, a middle power, by two bridges, and is within sight of Indonesia, one of the globe's weaker (but non-failed) states."[12]

Singapore is a city-state. The country's position at the southern extremity of the Strait of Malacca has helped it develop as a focal point for global shipping, both as a hub for oil refining and as one of the two busiest container ports in the world. More than 100,000 ships visit Singapore annually, averaging one vessel entering the Singapore Strait every four minutes.[13] And while Singapore dominates financially, Malaysia and Indonesia dominate geographically. Two noncontiguous halves principally form Malaysia, one on the peninsula north of Singapore, the other on the island of Borneo, which it shares with Indonesia (and Brunei). Indonesia is far more fractious by comparison. It is an archipelagic state composed of more than 17,000 islands spread across three million square kilometers of territorial and archipelagic waters, in addition to another three million square kilometers of exclusive economic zone.[14] Beyond Indonesia's thirty-nine legally recognized ports, the country's archipelagic nature has bred thousands of illegal ports, called "jalan tikus [alternative routes]," according to Brig. Gen. Dedy Fauzi Elhakim, an Indonesian counternarcotics officer.[15] The gradations in capacity and priorities among the three littoral states and neighbors including the Philippines and Thailand make for a compelling environment in which to test our broken windows best practices.

In another distinction from the previous case studies, Southeast Asia is wracked by more obvious great-power conflicts than are visible in the Caribbean or West Africa. These tensions promulgate a variety of significant threat scenarios of which naval planners must be cognizant, as noted by Sam Bateman, professor at the University of Wollongong's Australian National Centre for Ocean Resources and Security.[16] One of these major threats includes the potential for sovereignty skirmishes. The conditions that make such skirmishes possible, born of regional territorial disputes to which I will return, serve as obstacles to greater intraregional maritime cooperation. Mutual hostilities reign, particularly with respect to Singapore's perceived (contradictory) character as both an apparent Western and Chinese outpost. Such skirmishes could even precipitate what Bateman labels "catastrophic threats," such as a conflict between the Association of Southeast Asian Nations (ASEAN) states and China, perhaps over the disputed Spratly Islands, around which the straits would serve as the primary transoceanic chokepoint. Yet, as Bateman also notes, the probability of an incident occurring decreases as its lethality increases, meaning that the most endemic threats—namely, the small-scale "quality-of-life" crimes I have discussed at length—are not the ones about which strategists think most. Bateman identifies many of the same low-level issues as I described in the Caribbean Basin and the Gulf of Guinea, including pollution; piracy and robbery at sea; trafficking in goods and people; illegal, unreported, and unregulated (IUU) fishing; and climate change.[17]

Just as in the Caribbean and the Gulf of Guinea, narcotics trafficking and, consequently, money laundering are endemic. Two-thirds of the world's opium comes from South Asia and Southeast Asia, as noted by Aditi Chatterjee of New Delhi's National Maritime Foundation, precipitating threats to good order at sea.[18] Much of the world's illicit poppy was cultivated in the Golden Triangle region adjacent to these busy waterways, in Myanmar, Thailand, and Laos, drawing comparisons between Southeast Asia and the Caribbean. And while production numbers have fallen in recent years, as we saw in the Caribbean, drug trafficking is highly elastic. In another similarity to the Caribbean, maritime trafficking routes in Southeast Asia provide redundant and secretive alternatives to land-based paths. Alan Dupont, former director of the Asia-Pacific Security Program at the Australian National University, notes that predominant routes include traveling from Myanmar through seaports to Singapore and Malaysia or farther east departing from Thai coasts.[19] From Singapore and Malaysia, narcotics have been shipped to destinations as far as Australia and New Zealand. Amphetamines also built momentum in the late 1990s and early 2000s, moving between the Philippines, Malaysia, Indonesia, and Thailand, where the use of such drugs reached "epidemic proportions."[20] The impact of this trade is the same the world over. As seen across the previous chapters, "the networks and illicit markets supporting traffickers undermine the rule of law and weaken public institutions. They perpetuate corruption and contribute to geopolitical tensions throughout the region."[21]

Similarly, IUU fishing, as in the Gulf of Guinea, threatens the livelihood of regional subsistence fishermen, some of whom are catching just 15 percent of what they did a half-century ago, according to Bateman.[22] An estimate by the Indonesian government placed annual losses to IUU fishing at $4 billion, according to Middlebury College's David Rosenberg, much higher than the estimated cost of regional piracy.[23] For this reason, the Indonesian government has expanded a policy of sinking captured illegal trawlers.[24] Clashes among law enforcement (or proxy forces acting in lieu of law enforcement) and illegal (and legal) fishermen are frequent, occasionally violent, and run the risk of escalating tensions among claimants to various territorial disputes. The region is in fact replete with unresolved border disputes, including (per Bateman) a dispute over Malaysia and Indonesia's exclusive economic zones at the northern end of the Malacca Straits and sovereignty disputes between Malaysia and Singapore by the Singapore Strait.[25] Indonesia and Malaysia took a dispute over two islands near Borneo to the International Court of Justice, which, in 2002, ruled in Malaysia's favor. China, with the Nine-Dash Line map through which the nation lays claim to a substantial portion of the South China Sea, is a party to a great number of border disputes throughout the region. Consequentially, nearly all shipping transiting the Malacca and Sunda Straits must pass the Spratly Islands, as scholar Ashley Roach notes, and some of the most contentiously disputed real estate in the world.[26]

Like narcotics and IUU fishing, irregular migration and human trafficking are also abundantly present in the region. Extensive international coverage of the plight of stranded Rohingya Muslims, abandoned by their traffickers at sea, brought global attention to a long-standing issue in the region. Migrants in the region take various routes, many beginning on foot traveling through Thailand overland to Malaysia, although maritime routes persist as well (either over riverways or across the Andaman Sea). Indonesia and Malaysia serve as gateways for the smuggling and trafficking of people from as far afield as Afghanistan and Iraq into Australia, as Chatterjee writes, in everything from fishing boats to container ships.[27] Bradford notes that the associated criminality from this trafficking has pervasive effects on a population, including a proliferation of disease and communal violence (some of the very factors that precipitate migration flows in the first place).[28] Bateman, moreover, notes that greater poverty and climate change could provoke still greater human migration, much as discussed in chapters 1, 4, and 6.[29] And, as usual, corruption plays a critical role in greasing the wheels for the myriad maritime crimes committed in Southeast Asian waters, from "fees" paid by illegal fishermen to Indonesian naval personnel, to the involvement of some of the same personnel in human trafficking.[30] None of these issues can be resolved simply or in isolation.

Finally, as in the Gulf of Guinea, the maritime crime for which Southeast Asia is most commonly associated is piracy.[31] The region's geography plays a large role in the growth of this predation. As Chatterjee underscores, chokepoints like the straits force vessels to sharply reduce their speeds, making for easier targets.[32] High ship volumes (more than 70,000 a year), combined with natural hazards such as shallow reefs and numerous small islands, further force ships to reduce speed when entering the straits.[33] The narrowness of the waterways leaves slower ships vulnerable to attacks launched from shore, as Hastings notes, a tactical distinction from Somali piracy.[34] This geography was so favorable for maritime criminals that, by 2000, International Maritime Bureau (IMB) data held that the Strait of Malacca was the world's piracy epicenter. Meanwhile, Lloyd's, the insurance giant, labeled the region a de facto war zone in response to rising concerns over the nexus between piracy and terrorism.[35] As in the Caribbean and the Gulf of Guinea, criminals manipulate borders and geography while engaging in what Hastings calls "state capacity arbitrage," leveraging the strengths and weaknesses of neighboring countries against one another (i.e., stealing goods in an unsecured Malaysian port and laundering the proceeds through Singapore's modern financial infrastructure) and evading pursuit by crossing maritime borders.

By the end of the 1990s Southeast Asia had the highest piracy rates in the world—and they seemed to only be on the rise. From 1997 to 2002, attacks increased from 22 to 164, according to Adam Young and Mark Valencia (both at the East-West Center).[36] (My consolidation of IMB figures from a wider swath of Southeast Asia shows numbers rising that same period from 92 to 253, with the regional high-water mark coming the following year at 170.) Moreover, Bradford describes how, as pirates became better resourced and organized, the sophistication of attacks grew from hit-and-runs to hijackings and kidnappings.[37] The shipping community was quick to respond, establishing in 1992 the IMB's Piracy Reporting Centre (IMB-PRC) in Kuala Lumpur.[38] (Piracy imposes a number of costs on shipping. Beyond the value of goods stolen and rising insurance premiums, the need for tankers to sail at higher speeds to avoid being targeted could add almost $90,000 a day per ship in added fuel consumption costs.[39]) The international community lagged only slightly in comparison, taking an interest in the latter half of the decade, according to the Joint Services Command and Staff College's Catherine Zara Raymond.[40] According to Raymond, piracy in the region built more momentum as coastal communities wrestled with the Asian financial crisis. Several high-profile attacks from 1997 through the end of the decade may have also encouraged shippers to report attempted attacks, she notes, causing reports to rise from virtually zero to seventy-five in the span of only a few years. This history serves as a backdrop to a region-wide effort to combat piracy. As we will see, that response's initial successes and its long-term struggles offer a clear strategic case for the utility of a broken windows lens for maritime security.

Piracy

From 1994 through 2006, and then again from 2012 through at least 2016, the plurality of piracy reported worldwide has taken place in and around the South China Sea, clustered around Malaysia, Singapore, and, predominantly, Indonesia. Using IMB data, I calculate that 63 percent (1,607 out of 2,544) of all Southeast Asian piracy attempts took place off Indonesian waters alone (taking an expansive view of the region, including Brunei, Cambodia, Indonesia, Malacca, Singapore, Myanmar, the Philippines, Thailand, and Vietnam). Notwithstanding Indonesia's heavy footprint, however, piracy by its nature remains an international problem around the straits. For Singapore, whose financial health depends on the free flow of goods through the Straits of Malacca and Singapore, piracy has been a driving priority from the earliest signs of trouble, prompting investment in coastal surveillance and patrols from the 1990s. Malaysia soon followed suit, as did Indonesia under pressure from the other two littoral states, and counterpiracy became akin to a regional specialty.[41] While initially promising, these individual state efforts and nascent bilateral initiatives proved insubstantial in the face of mounting attacks, and the three states fell behind a rapid growth of piracy in the straits. As the international community took notice as the millennium came to a close, counterpiracy in the Straits of Malacca and Singapore entered a phase of investment and scrutiny that would prove fruitful, if perhaps unsustainable.

Response

Several diverse parties helped drive the littoral states toward an investment in maritime security. First, the growing impact of piracy on international shipping brought the IMB, the International Maritime Organization, shipping companies, and insurance companies to the table. Second, following the attacks of September 11, 2001, the United States took a greater interest in securing the straits from terrorism, an interest heightened by rumors circulating that pirates were hijacking ships in the straits in order to learn how to navigate (seen in the guise of the 9/11 hijackers attending flight schools). When, in June 2005, the Joint War Committee of insurance giant Lloyd's listed the Strait of Malacca as a war-risk zone, raising the cost of shipment for producers and user states alike, it became financially imperative for Singapore, and to a lesser extent Malaysia and Indonesia, to respond emphatically to the piracy epidemic.[42]

The regional response was robust and international at the highest reaches of government. As early as 2003, Roach notes that the members of the ASEAN Regional Forum addressed the threat of piracy at a ministerial level, through a declaration of cooperation on maritime security. In 2004 the littoral states initiated trilateral coordinated patrols under the operational name Malsindo to better organize their independent surveillance activities (Thailand became a partner of Operation Malsindo in 2008).[43] These patrols produced no immediate impact on piracy rates, Raymond records, although they clearly signaled a willingness on the part of regional states to act collectively on maritime security. One year later, the littoral states introduced joint air patrols under the Eyes in the Sky program, and in 2006 the two programs were merged into the Malacca Straits Sea Patrols.[44] Also in 2006 the Regional Cooperation Agreement on Combating Piracy and Armed Robbery against Ships in Asia (ReCAAP) entered into force. The agreement is another noteworthy example of successful multilateralism in the region—ReCAAP was the world's first intergovernmental organization devoted to countering piracy, according to Bradford[45]—and was established to facilitate information sharing, both regarding incidents and best practices.[46]

The importance of the free and financially expedient flow of goods through the straits weighs not only on the littoral states but on user states as well. One of the largest user states is Japan, which has been an enthusiastic contributor to stabilization efforts. Since the 1990s the maritime authorities of Southeast Asian nations have been invited to Japan to train with the Japan Coast Guard.[47] Roach writes that Japan's coast guard makes frequent visits across the region, offering instruction and collaboration in addition to the country's founding role in ReCAAP. With Japanese help, as Arizona State University's Sheldon Simon writes, Jakarta instituted a coast guard to augment and synthesize the disparate enforcement responsibilities of the nine agencies and local jurisdictions that owned authority over maritime enforcement.[48] Indonesia, as noted by Yann-huei Song of Taiwan's Academia Sinica, has made strides to unify this structure, creating navy control command centers and establishing regencies along international borders in an attempt to address piracy's perceived root causes, such as poverty.[49] Indonesia's Ministry of Home Affairs, meanwhile, has initiated similar "dissuasion programs," whose name at least reflects an implicit understanding of the importance of perception in maritime crime.[50] Malaysia has also attempted to centralize its maritime components under a unifying coast guard, the Malaysian Maritime Law Enforcement Agency, which reorganized five agencies under one roof in 2005.[51] Simon notes that Malaysia also invited the Japan Coast Guard to train personnel in that newly formed coast guard.[52] (Singapore, by contrast, has historically maintained a highly structured approach to maritime security, with a maritime and port authority, a police coast guard, and a navy; a maritime security task force coordinates their collective maritime efforts.[53]) Song notes that the Japanese nonprofit Nippon Foundation donated training vessels to Malaysia in 2006.[54] That same year the Japanese government announced it would donate several patrol boats to Indonesia. Malaysia has also been investing in capacity and surveillance independently. In 2000 the nation stood up a counterpiracy task force within the Royal Malaysian Marines and trained dozens of officers in a special "marine police tactical unit."[55] In 2003, Song adds, they acquired new patrol vessels and invested in radar stations to more effectively monitor traffic along the Malacca Strait. In 2006 Malaysia announced it would acquire up to fifteen new police vessels to aid in its goal of twenty-four-hour surveillance of the strait, and the country's navy acquired an auxiliary vessel to support maritime security operations in January 2016.[56]

The United States has also played a role in response. Simon notes the U.S. use of train-and-equip initiatives to build capacity in the region,[57] while Khalid mentions the role of funding lent to counterterror efforts like the South East Asian Regional Center for Counter-Terrorism.[58] Through the installation of coastal radars, ship transfers, and multinational exercises, the U.S. Navy and Coast Guard have delivered further millions in aid and training to the three littoral states.[59] The Cooperation and Readiness Afloat program is one such example, through which U.S. maritime services annually exercise alongside the three littoral states as well as select others (like Thailand and the Philippines). Southeast Asian Cooperation Against Terrorism is yet another U.S.-led multination maritime exercise in the region, begun in 2002. The United States has also donated matériel, including fifteen patrol boats, to the Indonesian Marine Police.[60] In 2011, for example, a retired Coast Guard cutter, USCGC *Hamilton* (WHEC 715), was transferred to the Philippine Navy and became BRP *Gregorio del Pilar* (PF-15).

Port visits have also increased over the past decade as emphasis on the region has grown, with U.S. Navy visits to Malaysia nearly tripling toward the end of the 2000s, and a large naval delegation (including the chief of naval operations and the aircraft carrier USS *George Washington* [CVN 73]) attending Indonesia's independence celebrations in 2009.[61] Also in 2009 the U.S. Navy established "formal, recurring talks" with Japan, South Korea, and Australia—all prominent user states; initiated high-level talks with Singapore; and expanded exchanges between fleet and combatant commanders and their regional counterparts to build personal and institutional relationships across services.[62] Yet another program, the Maritime Law Enforcement Initiative, builds more directly on the constabulary nature of maritime insecurity in the region, providing resources to help regional states combat human trafficking, narcotics and weapons smuggling, IUU fishing, and environmental crimes.[63]

Finally, intergovernmental and nongovernmental organizations have also lent a hand. In addition to the IMB's Piracy Reporting Centre, the International Maritime Organization has introduced programs like the Marine Electronic Highway project to enhance safe passage of the waterways.[64] Other international organizations involved in maritime security in Southeast Asia include ASEAN (including the ASEAN Regional Forum and the ASEAN Security Community), Asia-Pacific Economic Cooperation, and the World Bank, among others.

By the mid-2000s, piracy was waning across Southeast Asia. Figures peaked at 170 attacks in 2003, according to the IMB-PRC, and declined to a low of 46 in 2009. More specifically, figures within the straits declined from 80 reported in 2000 to 10 in 2007. Something, it seemed, was working. The literature, however, is inconclusive regarding which initiatives were most responsible for this overall decline in piracy rates. This is a critical point. Despite apparent downward trends in piracy rates, the literature remains largely unable to identify specific actions or programs responsible for the decline. "Probably," "may have," and "most likely" are often the closest that analyses come to linking the decline in reported incidents through the mid-2000s to the initiatives under way at the time. This is entirely forgivable; counterpiracy operations were not laboratory experiments. As in criminology, directly linking enforcement techniques with observed outcomes is fraught with difficulty. Nevertheless, we should remain skeptical about the degree to which specific efforts drove down piracy rates. Bradford notes, for example, that four years after the initiation of trilateral surface patrols among the littoral states, it remained unclear whether such actions shared any relationship with declines in incident rates.[65]

There are even a few competing explanations for the piracy decline that discount state action entirely. Perhaps the most common dissenting theory is that the 2004 tsunami that ravaged the region played a role in the reduction in piracy, although this is perhaps undercut by the reality that rates fell in regions untouched by the disaster as well.[66] Raymond posits a theory that the devastation wrought by the tsunami on piracy-prone Aceh, an Indonesian province long plagued by a separatist insurgency and hit hard by the surge, may have contributed to ending the conflict and, by extension, piracy in the broader region. The contagious and multidimensional nature of piracy depicted in chapter 6 and the tipping point argument explored in chapter 2 would certainly lend credence to an argument that a reduction in piracy in a particular hot spot could radiate effects outward.

Moreover, not everyone interprets the decline in regional piracy as a blanket success. Simon ponders whether piracy had disappeared in the straits or "simply relocated to the South China Sea and the Indian Ocean approaches to the strait," as the balloon phenomenon from part 2 would predict.[67] Bradford similarly notes that piracy rates did not decline uniformly, and that rates increased near Sabah, driving Malaysian piracy rates up threefold at the same time they were declining more broadly.[68] Simon notes surges in 2008 in the Riau Archipelago south of Singapore as well as in the region of the northern Malacca Strait, between Sumatra and Malaysia's west coast.[69] Piracy suppression in the Malacca Strait may have driven piracy eastward to the Singapore Strait and deeper into Indonesian waters, but it has had impacts further afield as well.[70] Raymond notes that while the international community was focused on reducing piracy incidents in the Strait of Malacca, insecurity grew in other waters connecting Indonesia, Malaysia, and the Philippines, including the Sulu and Celebes Seas.[71] These seas create a "triborder sea area" linking the Philippines, Malaysia, and Indonesia, a critical junction made a haven by transnational criminals.[72] Despite some well-founded cautions, however, the general consensus has been that a combination of the efforts noted here have had a collective impact on the issue of piracy. And they may have—for a time.

Resurgence

Until maritime insecurity is addressed in a broader context, states can only hope to cope with piracy. By dint of the frequent focus on a single threat vector, the counterpiracy suppression initiatives employed in Southeast Asia have generally failed to target the "architectures of intertwined security relationships" (multidimensionality) that make transnational crime so resilient and to which I have dedicated so much space in this book.[73] Lessons learned from the application of broken windows suggest that this narrow lens also ignores the driving effect of the environment on decision making and disorder. The rise in regional piracy rates in recent years suggests that piracy suppression is no longer adequate in the face of wider maritime insecurity. Indeed, after an interlude of rampant growth off the Horn of Africa, the plurality of total piracy has, since 2012, drifted across the Indian Ocean back to the Straits of Malacca and Singapore where the issue has thus far proven endemic—despite laudable efforts.

After decades of investment, the IMB continues to issue piracy warnings for mariners transiting near Indonesia, the Malacca Strait, and the Singapore Strait. And while rates have stayed low in some places, they have climbed in others, such as near Malaysia's Tanjung Piai cape and

Indonesia's Bintan Island, adjacent to Singapore.[74] Overall, reports of piracy in the region tripled from a trough in 2009 to a recent high in 2015, by which time half of the world's reported piracy took place off the coasts of the three littoral states. According to ReCAAP, 2014 was among the most dangerous years in nearly a decade.[75] And while the numbers from that year are not the largest on record, as a share of piracy incidents worldwide, 2014 and 2015 are so far the high-water marks for Southeast Asia (representing 58 and 60 percent of all attacks, respectively). Of the six countries that accounted for three-quarters of the world's reported piracy incidents in 2014, only Nigeria was outside of the Indian Ocean, with Singapore, Indonesia, and Malaysia comprising more than half of the attacks (according to the 2015 IMB report). Of the 36 "serious attacks" listed in the IMB's report, 20 took place in Southeast Asia, and almost half occurred near the three littoral states alone. Worldwide, of the ports that reported 3 or more incidents, half were within the three littoral states—5 in Indonesia, 1 in Malaysia, and 1 in the Singapore Straits. The total number of incidents that took place within these anchorages constituted two-thirds of attacks in ports. Similarly, of the 204 actual reported attacks in 2014, nearly two-thirds took place in the straits or waters adjacent to the littoral states.[76] Data from the first half of 2015 showed a continuation of these trends, as Southeast Asia—particularly the straits—struggled with a nearly 20 percent increase in reported piracy.[77] In 2016 IMB data painted a different picture globally, with a large (but not unprecedented) decline in Indonesia, even as the country remained the single largest victim of pirate attacks worldwide.[78] It is too soon to tell whether this presents a new era in regional piracy or, as seems more in kind with trends from the past twenty-five years, a periodic but temporary lull.

Even with global piracy numbers leveling out (in 2017) after the surge in Somali piracy, piracy in Southeast Asian persists. Pointedly, as overall incident rates (both attempted or successful) declined globally in the mid-2010s, the number of successful attacks jumped, due predominantly to incidents in Southeast Asian waters.[79] Rates have also proven highly erratic, with some year-on-year movements painting an alarmingly elastic picture. Piracy in Indonesia in 2016 dropped by half from the previous year, for example. Yet, in the past twenty years, piracy off that country's coast has oscillated between double- and triple-digit numbers *at least six times*. Malaysia, for its part, witnessed one of the largest year-on-year increases in incident reports in 2014, with moderate reversions following in 2015 and 2016.

Regardless of frequency, most of these attacks are far more reminiscent of the small-scale attacks described in the Gulf of Guinea than the sensational Somali model. As one news report remarked: "If Somali pirates act like muggers, attacking isolated targets out of sight of the authorities, south Asia's pirates act more like pick-pockets, using the crowd itself as cover."[80] Crowded waterways conceal criminals, as they did for the terrorists coming ashore in Mumbai in 2008, the scene with which I opened this book. This cover is part of the environment within which maritime disorder persists and cannot be disassociated from our understanding of security in the littorals. Similarly, as ReCAAP's assistant director, Lee Yin Hui, notes, more than half of the incidents reported are petty thefts, opportunistic in nature.[81] This type of piracy, reminiscent of the low-level crimes explored in chapter 2, could be particularly vulnerable to enforcement strategies informed by criminality.

Obstacles to Collective Effort

Even a weak state, with help, can garner investment to meet a specific threat in a specific location for a defined duration, as was the case in 2005 and 2006 when the littoral states were induced to action by Lloyd's war-risk zone designation.[82] Indonesia, for example, initiated Operation Gurita in 2005, deploying at least twenty ships to piracy-prone waters, according to Bradford.[83] Such surges are often unsustainable, however. In 2005, after at least four years (nonconsecutive) of triple-digit pirate incidents near Indonesian waters, as little as 25 to 30 percent of Indonesia's naval vessels were seaworthy, and the country maintained as few as one or two functional aircraft for maritime surveillance.[84] Indonesia's defense minister later put operational estimates higher but at a still-dispiriting 60 percent, according to Republic of Singapore Navy major Victor Huang. Even at full operational levels, the Indonesian fleet is less than half the size necessary to patrol its waters, according to estimates of a former naval chief of staff relayed by Huang.[85] And as noted by the former head of the Piracy Reporting Centre, Indonesia's investments really matter: "A lot lies with Indonesia and with how long they will sustain their patrols, which do cost millions."[86] Indonesia's "anemic maritime budget," in Simon's assessment, makes it impossible for Jakarta to maintain situational awareness over its 17,000 islands.[87] Complicity in local crimes among elements of the Indonesian Navy and maritime police exacerbates cultural and systemic problems in much the way corruption does in the Caribbean. Some peripheral states are no less well off. In late 2014 the Philippine Navy announced plans to modernize a fleet that, until then, had been marked predominantly by "museum pieces."[88] To compensate, Simon writes that national authorities have implored local fishers to report the maritime crimes they see in their coastal communities, instituting (perhaps inadvertently) the central element of any good broken windows community policing approach.[89] As we saw in previous chapters, this community engagement can have a profound impact. In part 2, locals in Colombia detected and disrupted a narco-sub production ring in partnership with local authorities. Meanwhile, Jamaican fishers served as bellwethers of changing trafficking routes, sensing increased risks despite limited interdiction figures on the island. In the Gulf of Guinea, fishers from a Senegalese village reported a suspicious vessel to the local gendarmerie, who found it packed with cocaine. Trust between communities and authorities along with a context that engenders a sense of ownership for one's neighborhood produces actionable outcomes that increase local and transnational security.

Yet there is more to the shortfalls of maritime security in the Straits of Malacca and Singapore than resource constraints. A range of issues, including sovereignty sensitivities and issue prioritization, serve as obstacles to the collective framing and action necessary to make the midcentury policing model (interdiction based) effective in the maritime space. As mentioned earlier, territorial claims in Southeast Asia bracket any conversation on regional security frameworks. Sovereignty concerns are a recurring refrain in the literature and are especially important for Malaysia and Indonesia. This emphasis on territorial integrity is an outgrowth of the noncontiguous, archipelagic, and maritime nature of these two states, whose wealth and influence are directly related to the extent of their dominion over adjacent waterways. Due to such sovereignty concerns, Indonesia and Malaysia have been slow to engage in multilateral pacts beyond the three littoral states. They are not, for example, signatories to the Convention for the Suppression of Unlawful Acts, nor are they party to ReCAAP.[90] Even Indonesia, most in need of capacity building aid, has drawn a red line around the presence of foreign militaries, according to Huang.[91]

When, in 1999, Japan's prime minister, Keizō Obuchi, proposed a regional coast guard initiative, ASEAN states responded tepidly and China with outright opposition.[92] While some of this tepidity is a result of Japan's wartime legacy in East Asia, modern politics plays a role as well. China, as interpreted by Ralf Emmers (at the time a postdoc at the Institute of Defence and Strategic Studies at Singapore's Nanyang Technological University), viewed the Japanese initiative as an avenue through which to expand regional influence. Indonesia and Malaysia went so far as to agree to participate in exercises but balked at the prospect of armed vessels in their territorial waters, Emmers writes.[93] An inadequate appreciation for the depths of sovereignty considerations was also a factor in the defeat of an earlier Japanese proposal in 1996, according to Bradford.[94] As a result of these defeats,

Huang concludes that, "in general, attempts by extraregional powers to exert leadership are likely to trigger unfavorable reactions from rivals."[95]

American initiatives are viewed no less suspiciously by the littoral states. In March 2004 misreported (or misinterpreted) remarks by Adm. Thomas Fargo, who was quoted as suggesting the presence of American sailors in the straits under the RMSI, prompted quick rebuke. Malaysian prime minister Abdullah Ahmad Badawi quipped, "I think we can look after our own area."[96] U.S. ambassador to Indonesia Ralph Boyce was later forced to issue a clarification, stating that the news reports were inaccurate.[97] One Malaysian maritime security commentator, relayed by Song, considered the perceived American attempt to send troops to the region under RMSI as a Singaporean ploy intended to thwart the modernizing Malaysian military. Another expert Song identifies as a former director general of the Maritime Institute of Malaysia wondered if RMSI disguised an American effort to seek out new enemies in the region, "using the Straits of Malacca as a pretext" or concealed a "hidden agenda." Song catalogues several elements of the RMSI program that rankled regional officials, including familiar references to sovereignty and political spillover as well as doubts of American sincerity and intentions. The overt regional hostility to RMSI's perceived intentions demonstrates this tendency toward "ambivalence or outright rejection" (Huang's words) of most frameworks introduced by extraregional powers.[98] Any successful vision for maritime security in the region must negotiate the need to balance external funding and interests with a perspective on security that recognizes local territorial and political concerns.

Bateman makes a similar argument, writing that "a fixation on resolving sovereignty claims reinforces distrust and inhibits cooperation, whereas cooperative actions on 'softer' issues should be seen as CBMs [confidence building measures] helping build trust."[99] And as I have argued throughout this book, a cooperative approach is enabled by (and benefits from) a theory of maritime security that is not zero-sum (i.e., we address one issue at the expense of not addressing another). Without developing a theory of maritime security that allows the region to move outside of sovereignty clashes and build cooperation, the myriad quality-of-life crimes (Bateman's "softer" issues) that prey on lives and livelihoods in the region will continue to proliferate. This was what was on display in the regional response to RMSI. According to some reports, the Indonesian administration understood that Admiral Fargo's comments were misquoted but nevertheless opposed the program because it was "*perceived* as overly militaristic" (emphasis added).[100] Successful maritime security cooperation in the region may depend on the capacity to frame disorder in less militaristic, zero-sum, even counterterror terms and more in ways that emphasize communal security and law enforcement.

As we saw in discussions on both the Caribbean and the Gulf of Guinea, attempting to coordinate a vast array of governments and intergovernmental organizations invested in maritime security is a daunting challenge. On June 14, 2014, the tanker *Ai Maru* sent a distress signal to the IMB-PRC. Six ships from four agencies belonging to three countries were dispatched to respond to the call, and the pirates still evaded capture.[101] The mere presence of international coordination does not preordain its success. It should come as no surprise, therefore, that those international agreements that *have* entered force have often been viewed as disjointed or ineffectual. ReCAAP, for all its accomplishments, remains "incapable of having a large impact," according to University of Wollongong's Ahmad Amri.[102] Amri is even less optimistic about the impact of ASEAN's forums, calling them "merely talk shops." Critics have also seen the cooperative sea and air patrols discussed above as "more show than substance," in Bradford's words.[103] As many as seventy air sorties a week are estimated as necessary to effectively monitor the waterways, while Eyes in the Sky generally executed only eight, lending further justification to this criticism (and questions over the causality of this enforcement on piracy declines).[104] Moreover, the degree of territory in need of coverage and the paucity of resources with which to do so are reminiscent of similar problems in the Caribbean, which earlier demonstrated the inadequacies of a midcentury-policing-styled interdiction model in the littorals.

Most importantly, however, the littoral states failed to maintain a parallel level of effort in the fight against piracy because piracy is simply not a priority for all parties. Malaysia, for example, whose Maritime Enforcement Agency maintains a respectable seventy patrol ships, regards illegal fishing and tourism development as higher priorities than piracy, according to both Simon and Major Huang.[105] Meanwhile, as both authors note, Indonesia does not face the same financial incentives that motivate Singapore's interest in piracy suppression. While Indonesian goods principally traverse alternative straits, like Lombok and Makassar, IMB data from 2015 show that Singaporean-flagged or -managed ships were among the most victimized near the height of the region's troubles in 2014. Moreover, Jakarta maintains a long list of nautical issues of concern, including maritime territorial disputes, smuggling, and IUU fishing, not to mention landward sovereignty—all of which rank higher in the interest of the state than counterpiracy. Both Malaysia and Indonesia regard human trafficking as a major issue as well. Singapore, by contrast, is far more concerned with hazards to the free flow of traffic through its adjacent straits, as Simon is right to note, a symptom of the city-state's heavy dependence on the shipping industry. Simon likewise points out that even user states have conflicting priorities and perceptions. The United States, as we will see, has been predominantly concerned with the threat of terrorism, Japan with piracy, and Australia with overall capacity building (given its role as a destination for smuggled products and people).[106] That the range of maritime crimes represented in these seemingly divergent interests is interconnected, as we have seen, is diminished in a strategic lexicon pitted against individual threats. These barriers to collective action—resource allocation, concerns over sovereignty, and mismatched threat perceptions—are illustrative of the pitfalls of strategies that do not consider the broader context of maritime insecurity.

The Straits of Malacca and Singapore represent a unique case in this study. For a time, efforts predicated on traditional notions of threat framing seemed to be making a real impact. Yet what makes this case all the more powerful is that even promising counterpiracy efforts do not offer the theoretical frameworks that are critical to sustaining "resolve and resources" over the long run. Bradford argues that, should the littoral states manage to succeed in maintaining momentum and resource investment in counterpiracy, this success can be parlayed into a mechanism to drive cooperation on a wider set of maritime insecurities in the region.[107] As I have argued throughout these pages, however, the reverse may be a more efficacious approach. Buy-in can be elicited by developing a strategic approach that addresses a host of insecurities, placing linkages between these threats and the relationship between security and context at the forefront. This multifaceted focus would enable regional partners to sell the approach politically, given that priorities differ across actors, while also targeting on disassembling the infrastructure that undergirds a diverse set of crimes. Such a strategy's law enforcement heritage may be able to avoid the stereotyping of typical security regimes that "[set] off power rivalries and sovereignty concerns" in the region and could make real inroads in Southeast Asia.[108]

Obstacles to long-standing piracy reduction in the straits suggest that there is room for progress in the strategies brought to bear on regional maritime insecurity. As in the Caribbean and the Gulf of Guinea, sensitive political considerations, difficulties in international coordination, and the infeasibility of fully resourcing an interdiction-oriented approach in the Straits of Malacca and Singapore serve as further evidence that existing structures leave room for improvement. These findings are not new. Indeed, the literature has evolved considerably over the years as piracy proved an endemic concern. Yet, when we analyze the changing academic discourse on regional maritime security, we can see that a broken windows lens would add greater clarity and body to this evolution. While the literature lacks the framing and language of the broken windows theory, from which it would benefit, a manifest gravitation toward a multidimensional and contextual understanding of maritime security in the discourse is indicative of the theory's relevance even in Southeast Asia.

EVOLUTION TO BROKEN WINDOWS

Piracy's multidimensionality is increasingly taken as self-evident. Bradford explicitly connects piracy with terrorism, smuggling, and migration, for example. Likewise, references to the linkages among instability, corruption, poverty, and transnational crimes serve as an uncontroversial acknowledgment of the need for a more holistic interpretation of what maritime security means. The spike in academic interest in terrorism after 9/11 may have provided a catalyst for this increasingly nuanced view of piracy. As terrorism gained ground as a field of study, that literature inadvertently served as a vehicle by which that on Southeast Asian maritime security was exposed to a more multidimensional perspective. All of this, meanwhile, may have been precipitated in part by the shortcomings of the midcentury model as applied to piracy, which was resource intensive yet unable to fundamentally change the piracy calculus. In this section, I trace the literature's transition to further underscore the relevance and value of the broken windows lens in framing piracy and maritime security.

From Piracy to Terror

No work on maritime security in the straits is complete without reference to terrorism—often right in the title (of which I am surely guilty as well). The relationship between 9/11 and the maritime security literature is profound and difficult to overstate. Raymond, for example, writes simply: "One year after piracy incidents peaked in the Malacca Strait, al-Qa'ida launched its attack on the Twin Towers in New York."[109] Raymond was of course not arguing any correlation; rather, 9/11 demonstrated that conveyances (i.e., planes and, by extension, ships) were now seen as genuine vessels for terror attacks. Nevertheless, the acute juxtaposition is illustrative—and common. As I noted earlier, the literature draws frequent parallels between the flight school training of the 9/11 hijackers and concerns over similar activities in the straits. Reports of ship hijackings are often portrayed as potential "dry runs," as Roach writes, in which terrorists gain operational knowledge for later use—perhaps ramming an explosives-laden ship into a port—though some experts have disputed these conclusions.[110] One example portrayed in this light is the March 2003 hijacking of the chemical tanker *Dewi Madrim*, which, after being commandeered, was steered through the Malacca Strait for an hour before the assailants fled with a trove of technical documents.[111] (As noted by the University of St. Andrews' Peter Lehr in discussion with me, the ship's master later confirmed it was just a "run-of-the-mill" pirate attack.) Gal Luft and Anne Korin, of the Institute for the Analysis of Global Security, even relay reports that Al-Qaeda owns dozens of "phantom ships," hijacked and repainted vessels.[112] Young and Valencia posit a more exact figure—fifteen ships—and regard their existence as a "widely disseminated, and largely accepted, fact."[113]

The list of hypothetical scenarios involving merchant shipping disasters is extensive. Bateman, for example, sketches several theoretical catastrophic attacks, including exploding liquid natural gas tankers, ferry bombings, and cruise liner attacks.[114] Such acts of terrorism would have particularly adverse economic consequences for Singapore given the nation's reliance on an extensive, advanced coastal infrastructure, which Major Huang notes is itself an attractive target.[115] Should a tanker be sunk in a narrow portion of the straits, such as the Phillips Channel in the Singapore Strait, the cost in loss of business to Singaporean ports could amount to hundreds of millions of dollars annualized.[116] Regardless of the perhaps sensational nature of some scenarios, their underlying premises remain compelling. Moreover, given the presence of local extremist movements in Southeast Asia, concern of a more robust cross-pollination between piracy and terrorism is ever present. Many organizations in the region, including the Free Aceh Movement, the Abu Sayyaf Group, the Moro Islamic Liberation Front, and the Moro National Liberation Front, have all employed maritime terror tactics or engaged in piracy.[117]

Similar to the use of narcotics trafficking in funding terror in Latin America and the use of piracy in West Africa to fund hybrid political-criminal organizations, piracy in Southeast Asia has also been suspected as a source of funding for extremism. Senia Febrica, a researcher at the University of Indonesia, notes that terrorist organizations (including the Abu Sayyaf Group and the Moro Islamic Liberation Front) have been indicted for attacks in the adjacent Sulu Sea designed to provide their organizations with funding. Febrica also notes that sea passages between Indonesia and the Philippines have been used as conduits for terrorist movement, including the passage of individuals from Malaysia destined for training camps in the Philippines.[118] Luft and Korin relay that international efforts on countering terrorist financing have pushed terror organizations toward alternative funding streams like piracy, and that this is particularly evident in the Strait of Malacca.[119] As Chatterjee notes, narcotics and weapons trafficking fed from as far afield as Afghanistan to Thailand sustains forces of instability (like insurgents or terrorists) well across the Indian Ocean region, with some matériel arriving by way of fishing boats crisscrossing the Andaman Sea.[120] The movement of narcotics, as we have seen, also precipitates the involvement of more diverse criminal syndicates, including Japanese yakuza, Chinese triads, and Vietnamese gangs.[121] As Luft and Korin note, pirates also often have relationships with other transnational criminal organizations operating in Southeast Asian waters.[122]

The focus on terror is not merely academic. While incidents of maritime terrorism have been scarce in the waters around the South China Sea, they are not unheard of, often coming in the form of ferry bombings. In 2000 the Philippine ferry *Our Lady Mediatrix* was bombed, allegedly by the Moro Islamic Liberation Front, killing approximately forty people. In 2001 the Indonesian ferry *Kalifornia* was attacked, killing ten passengers.[123] In one of the most frequently cited examples in the literature, the 2004 attack on Superferry 14 in Manila Bay resulted in the most lives ever lost to an act of maritime terror; more than one hundred people died. Abu Sayyaf claimed responsibility for that bombing. All told, in the early 2000s, attacks on ferries in the region killed approximately three hundred people.[124]

States, too, have exhibited real concern over the issue. Singapore's former defense minister Teo Chee Hean has "consistently maintained" that maritime terrorism poses a real threat to the region, according to Huang. Malaysia, which has previously downplayed the threat of terrorism, also now warns of the threat in the region and has referred to it as "real and possible."[125] Indonesia, which has similarly demurred from drawing links between piracy and terrorism in the past, also began to make such linkages in the years after 9/11.[126] Indonesian officials have relayed intelligence that Jemaah Islamiyah (JI) has considered attacks on shipping passing through the straits.[127] Videotapes recovered from JI have included footage of Malaysian Maritime Law Enforcement Agency patrols, and JI militants have received training in maritime tactics, according to Song. Only months after 9/11, authorities uncovered a JI plot to attack American naval ships docked at Singapore's Changi Naval Base.[128] Captured Al-Qaeda member Omar al-Faruq has also reportedly acknowledged attempting to plan scuba attacks against American ships in the Indonesia port of Surabaya.[129] In 2010, Bradford points out, all three littoral states issued warnings of potential maritime attacks within the straits.[130] There have also been concerns that sympathizers from Southeast Asia were using sea routes to travel to and from Islamic State territory in the Middle East although an investigation by the Indonesian navy failed to turn up corroborating evidence. As the Islamic State crumbled in Iraq, a jihadist resurgence in the southern Philippines renewed concerns about terrorist use of Southeast Asian waterways yet again.

This attention to terrorism was also a major factor in the United States' interest in regional maritime security, as noted earlier. Song provides a litany of examples from the early 2000s.[131] In the months after 9/11, for instance, the United States collaborated with the Indian Navy to help protect American-flagged ships transiting the northern portion of the Strait of Malacca. Meanwhile, in 2004, the U.S. deputy assistant secretary of state for East Asia, Matthew Daley, told an audience at a dialogue on security in Asia that "Asia's waters are prime targets for Al-Qaeda and other terrorists," and that the threat "is not to be underestimated."[132] During Secretary of State Condoleezza Rice's visit to Indonesia in 2006, she discussed with her counterparts terrorism's interface with the security of the straits, and an American-Indonesian joint exercise on maritime counterterrorism took place shortly after Rice's departure.[133] American concerns over the safety of the straits became so critical so quickly that, by only 2002, the region had already earned the moniker "the second front" in counterterrorism by some observers.[134] The United States' focus on maritime terror has in part made it easier to partner with littoral states (on all manner of maritime issues), which have themselves come to see terrorism both as a regional problem and an avenue for accessing American resources.

As Young and Valencia rightfully address, the post-9/11 "conflation" of piracy and terrorism appeared ubiquitous and was undoubtedly overstated. Nevertheless, the commingling of terrorism and piracy in the Southeast Asian literature inadvertently marks an important theoretical step in conceptualizing maritime security in the straits. The predominance of terrorism in the approaches of policymakers and academics encouraged progress toward multidimensionality, blending two previously bifurcated policy spheres into a single threat picture. Within this framing, piracy and terror are still often regarded as "distinct crimes," according to Sittnick, and therefore treated as totally separate issues.[135] Still, as a consequence of their juxtaposition, piracy and terror come to be presented in the same context, manipulating the same vulnerabilities. While the motivations driving piracy and terror remain split when evaluated from this perspective—the former economic, the latter political—the actions of both organizations are seen to overlap, as are the environments in which they thrive.

Terror, Transnationalism, and Multidimensionality
Concurrent with this shift in emphasis toward terrorism and terror financing, in both the academic literature and international attention, is an expanding focus on other transnational crimes in the straits. In some cases, these

crimes are seen in similarly narrow lenses as piracy and terrorism, as isolated threats. Drug trafficking, for example, which Emmers notes came into greater focus as a critical issue for transshipment states as local consumption rose (like we saw in the Caribbean), is often discussed in isolation.[136] Yet the blending of terrorism, piracy, and transnational crime in the literature has also bred an increasingly multidimensional view of maritime security. This shift was facilitated by the global nature of Al-Qaeda's attacks in the early 2000s, which precipitated a vision of terrorism as thoroughly internationalized. Thus, while piracy was typically regarded as a matter for the littoral states to address among themselves, the integration of terrorism into the maritime threat landscape enabled regional states to more broadly consider nonterritorial, nonconventional threats. Febrica, for example, relays the common concern that irregular migration across porous maritime borders may be used to facilitate the easy movement of terrorists in the larger migrant flow.[137] Moreover, as Bradford notes, transnational criminal activity, particularly the movement of weapons and people, has a real impact on the capacity of nonstate actors to engage in political violence.[138] Luft and Korin posit that the maritime domain has even become a key operational realm for terrorists and that attacks at sea might represent more than merely attempts at financial gain.[139] Young and Valencia term this "political piracy," something that bares similarities to the hybridized maritime violence explored in chapter 6.[140] Song notes that the United States "sensed increasing synergy between transnational threats like terrorism, illicit drugs, trafficking in humans, [and] piracy."[141] Nazery Khalid, of the Maritime Institute of Malaysia, represents the evolution in the literature well. He notes that, despite a sustained media fascination with piracy, "the threats facing the Straits are multidimensional," and there are indeed "many other threats facing this key artery of world trade."[142] These conclusions are not revolutionary, but they demonstrate a progression toward a wider understanding of the maritime security domain. Now it is axiomatic that the threats of maritime security in the straits are "complex, asymmetrical, multidimensional and transnational."[143]

As we saw in previous chapters, it is a logical conceptual step from imagining terrorists and other transnational actors operating alongside one another to the fusion of actor and infrastructures. Chatterjee comes to this same conclusion, noting that "it is plausible that pirates and terrorists could collaborate" with one another, one gaining financially, the other logistically.[144] This leads to what Donna Nincic of the California Maritime Academy considers to be the most salient challenge facing the maritime domain, "the emerging 'web of criminality' with pirates, terrorists, and ordinary criminals working in an ad hoc manner."[145] Nincic refers to this expansion of piracy into a range of other maritime crimes as "diversion," or diversification, something I discuss at length in the previous chapters and something that is evocative of the multidimensional element of a broken windows' lens. A report from a maritime news service similarly suggests that Southeast Asian piracy groups are highly international in nature, with Indonesian foot soldiers and Malaysian and Singaporean bosses, and they often engage in a wide variety of illegal maritime activities.[146] As we have already explored, maritime criminals are more dynamic than the atomized state response suggests. As Bateman remarks, and as should be unsurprising by now, smuggling networks in Southeast Asia are elastic—the same gangs may be involved in smuggling one day and piracy the next.[147] Chatterjee writes of human traffickers that they are likewise connected to transnational networks engaging in the trade of narcotics, guns, money, and counterfeit products.[148] Capt. Richard Phillips, of the famed MV *Maersk Alabama*, notes that pirates operating in Southeast Asian waters are part of a larger criminal infrastructure.[149] Raymond, too, notes the significance not only of deterring pirates at sea but of tackling the broader network of criminality that sustains them.[150]

The inclusion of transnational crimes in the literature drastically alters the perception of the success of midcentury-style counterpiracy initiatives. As E. L. Dabova, of Saint Petersburg State University, notes, "While the actual number of piracy cases may be perceived as dropping, the total number of maritime crimes in the Strait of Malacca has actually increased."[151] Weighing heavily on this rise is the role of human smuggling and trafficking. Dabova estimates that a boat of between fifty and one hundred migrants could net traffickers as much as $30,000 while products from stolen motorcycles to timber all yield lucrative sums. The transportation of hazardous waste, one of the most visible symptoms of disorder, is among the fastest-growing crimes in the region.[152] Like the decentralized criminal syndicates that have been discussed, these trafficking groups are often small in size, between five and ten people (in Dabova's estimation). This localized, nodal view of crime is easily highlighted by the broken windows theory's community-oriented perspective but may be missed by a midcentury model focused on interdiction or the "decapitation" model used against more corporately structured criminal organizations. Due to the growing awareness of the multidimensional, contagious nature of littoral insecurity, issues that were once predominantly the concerns of one or two littoral states are now gaining traction internationally. Even less prominent crimes, like the smuggling of cigarettes, counterfeit medication, and even sand, are receiving increased focus because of their role as part of a larger system. The consequence of these crimes may, individually, be small. Yet, as Nincic argues, collectively they are "even more problematic because [they are] all part of the increasingly overlapping 'web of criminality' connected through the global maritime domain."[153] Just as George Kelling and James Wilson argued in 1982, so too does Khalid: disorder compounds and, if unattended, could have profound implications for the security of littoral states.[154]

HA/DR operations are also increasingly a part of the conversation. As Huang notes, the United States' role in humanitarian relief in the aftermath of the 2004 tsunami built operational capacity and political goodwill.[155] Bradford similarly notes that HA/DR operations performed in times of peace are integral to building the trust necessary for effective naval cooperation in times of crisis. Moreover, as broken windows would suggest, by aiding in recovery efforts and physically cleaning up after a natural disaster, nonmilitary actions can directly improve an environment and, thus, regional stability. The United States has embraced this premise in Southeast Asia. Since 2006 the United States has partnered with other Indo-Pacific nations in the Pacific Partnership initiative, providing humanitarian assistance across the region.[156] The hospital ship USNS *Mercy* (T-AH 19) has participated in the Sail Banda exercise alongside medical ships from Indonesia and Singapore.[157] As the author Robert Kaplan expresses, such deployments are likely to become increasingly common. Echoing sentiments from our exploration of population growth, climate change, and natural disasters in chapter 1, Kaplan writes, given "population growth in climatically and seismically fragile zones ... placing more human beings in danger's way than at almost any other time in history, one deployment will quickly follow another."[158]

Even the littoral combat ship (LCS), that much-maligned vessel, has made an appearance in the straits. The U.S. Navy has already operated several LCS from Singapore with an intention to station several forward in a more permanent capacity. Their presence, according to the commodore of LCS Squadron One, Capt. Randy Garner, offers "an amazing return on the shipbuilding dollar."[159] The ship's smaller stature allows it to better perform theater security cooperation in concert with the smaller ships of adjacent navies and coast guards. In a similar vein, Scott Cheney-Peters, formerly of the State Department's maritime security desk (and a retired U.S. Navy officer), notes that the "white hulls" of the U.S. Coast Guard are better suited still for the constabulary and criminal challenges faced by littoral states. And despite being short on assets, he writes, the Coast Guard does occasionally send vessels through the region to participate in joint exercises, as when (in 2012) USCGC *Waesche* (WMSL 751) participated in that year's Cooperation and Readiness Afloat operation. Cheney-Peters likewise notes that law enforcement detachments provide a unique solution for multiplying the effect of Coast Guard and law enforcement assets in the Straits of Malacca and Singapore, and that they "have a strong case to represent U.S. support" in the region, underscoring further the hybrid military-police dimension of maritime security needs.[160]

The special role of the Coast Guard in this sphere should not be understated. As Bateman notes, regional states are often concerned that the American interest in maritime security and counterterrorism has overly militarized their approach to the region and its strategy therein. Just as academics and practitioners in the Caribbean were concerned with a "fortress Caribbean" approach to security, leading to the urbicide we explored previously, so too are some concerned over a "fortress Malacca." The use of law enforcement detachments in the Caribbean as well as the Caribbean's gains in developing a cooperative theory of maritime security have even been explicitly held up as models for adoption in the straits. Through such a perspective, regional states can "build operational trust by disconnecting forms of maritime cooperation on the 'softer' issues" from the thornier geopolitical problems that drive long-held suspicions.[161] As a consequence, states would construct a more productive framework for engaging on maritime security, central to which is "demilitarizing" the way in which regional states approach issues of disorder at sea.[162] As we've seen throughout, there is a fine line to toe between addressing fundamentally destabilizing issues and doing so in a way that is somehow less than purely military. An approach that focuses on quality-of-life, community issues, like the broken windows theory, could provide the foundation to accomplish that.

This transition—from piracy, to terrorism, to transnational crime, to a multidimensional perspective on security—inevitably raises the question: how do states reconcile the criminal dimension of these issues with the resources necessary to combat them? That tension has been at the core of this book and is present at times in the literature on Southeast Asian maritime security. As Young and Valencia note, it is not the traditional responsibility of militaries to provide police functions.[163] While navies are often tasked to address such problems because they possess the needed hardware, they often lack the arrest authority and knowledge base necessary to police effectively. Yet the hybrid nature of crimes like piracy, terrorism, and trafficking in people, weapons, and goods is not easily addressed by conventional policing authorities either (recall the U.S. regulation of posse comitatus). As far back as 1999, Dupont wrote of narcotics trafficking in Asia: while the matter has been "traditionally regarded as principally a matter for law enforcement agencies," this is "demonstrably no longer the case."[164]

There is, similarly, an impulse to emphasize root causes, often with the assumption that short-term mitigation is unlikely to make a significant difference. Young and Valencia are emblematic of this argument, advocating for a strategy of maritime security that stresses addressing root causes rather than their symptoms (something we saw earlier in the chapter and across chapter 2). And undoubtedly, states should focus on governance, poverty alleviation, and human rights. Yet, as this investigation of the broken windows theory has shown, the symptoms matter as well. By tackling symptoms, we influence perceptions, which impact how people behave. As the implementation of broken windows policing in New York suggests, even without solving "root causes," maritime forces, "acting in a law enforcement capacity, can still have a significant effect."[165]

Still, one of the greatest endorsements of the need for an alternative way of looking at maritime threats in Southeast Asia comes from the limits of the dominant, midcentury policing-style strategy expounded over the years. The littoral states responded robustly to the rise of piracy in the region and have sustained a focus on maritime insecurity to varying degrees for more than twenty years. Yet, after a period of temporary piracy decline, renewed incident rates, and concerns over resources and matched level of effort, there is yet a more fundamental question about Southeast Asian maritime strategy. Are any of the lessons learned in the region, after two decades of counterpiracy operations, applicable elsewhere? Does the existing theory in the region pass the generalizability test I have striven to apply to my own? Simon, directing his evaluation toward the Gulf of Aden specifically, answers emphatically no: "any isomorphism seems remote" because of mismatches in geography, sociology and governance.[166] That after so many years the practice of maritime security in Southeast Asia may not have produced generalizable applications suggests the need for continued theorizing.

Finally, within piracy studies, some authors have even echoed some of the core sentiments of the broken windows theory, without the theory's wider scaffolding. Nincic, for example, suggests that the disorder brought along by piracy breeds greater disorder. Her depiction is reminiscent of that which I present in chapter 2, of a broken window's cascading impact on a local community. She writes, "Left unchecked, piracy tends to transition through levels of increasing complexity and violence. When it first appears, piracy tends to be little more than petty crime; left unchecked, it will tend to become more organized, moving from 'mom and pop' operations to sophisticated criminal gangs."[167] There is even reference to the contagious nature of piracy, something noted in the Gulf of Guinea study, akin to the vision of crime and disorder expressed by Kelling, Wilson, and Catherine Coles. Chatterjee even makes reference to the impact of context, identifying the region's "permissive environment" as a contributing factor to the proliferation of maritime crime.[168] References to the "synergetic" impact of transnational crimes can likewise be read as references to the compounding nature of disorder as understood in the lexicon of broken windows.[169] Robert Kaplan summarizes, "Piracy is the maritime ripple effect of land-based anarchy."[170] Broken windows substitutes the word "disorder" for "anarchy," but the principle remains. As one saying from the fifteenth century quips, "Whoever is lord of Malacca has his hand on the throat of Venice."[171] A key argument of this book, of the lessons gleaned from the broken windows theory and a point of convergence within the literature on Southeast Asia, is that "more safety and security in one of the world's busiest waterways constitutes a positive sum outcome" for everyone, as resonant today as ever, it would seem.[172]

The literature on Southeast Asian maritime security is among the most mature on the subject. The growth of the piracy epidemic in the 1990s and the intensity with which the global community demanded action have spurred an ongoing interest in maritime security research from Singapore to Australia. There is perhaps no greater endorsement of the potential usefulness of the broken windows theory than the growing prominence of our two main themes—multidimensionality and context-specificity—in this body of work. As Chatterjee relays, "There is a growing need to expand the traditional notion of security to address non-traditional threats, and thereby adopt a more comprehensive approach to security. . . . Apart from being non-military in nature, these challenges are also transnational."[173] Dabova likewise writes of the need to "develop a complex concept of non-traditional threat to maritime security" because current frameworks on security face "conceptual and institutional weakness."[174] The scholarship on Southeast Asian maritime security suggests simultaneously that there is an ongoing need to place maritime security strategy in a greater theoretical context and that the critical themes we have distilled from the broken windows theory meld with the lessons emerging in this robust literature. Thus, in demonstrating the applicability and viability of a broken windows–inspired strategy in a region with a long relationship with maritime insecurity—and piracy in particular—I underscore again the generalizability of this research.

Across every region, "policymakers have a hard time conceptualizing non-traditional, transnational and human security issues, which do not respect national boundaries and which transcend institutional and policy stovepipes."[175] In response, Nincic summarizes the need for this theorizing best: "The global maritime community should focus less on stopping a particular form of maritime crime . . . and more about creating holistic measures to counter an increasing 'web' of maritime criminality."[176] The broken windows theory, through an emphasis on multidimensionality and context, cuts across these stovepipes to present a holistic view of maritime security. Such a perspective presents more than just another way of looking at threats. As Richard Bitzinger, of Singapore's S. Rajaratnam School of International Studies, argues, "above all you need a common threat perception" to engender genuine cooperation and impact in the maritime domain.[177] A theory of maritime security, such as one predicated on the lens and lessons of the broken windows theory as explored in this book, provides one such avenue for redefining a shared threat perception.

❖ ❖ ❖

In this case, we examined a region that has directly addressed a prominent issue of maritime security with apparent success. Unlike the Caribbean or the Gulf of Guinea, the Straits of Malacca and Singapore are dominated by only a few states, some with real capacity for action in the maritime space. Despite this focus and latent capacity, we nevertheless saw that piracy remains nearly endemic. Through an investigation of piracy's resurgence in Southeast Asia, we saw in clear relief the continued need for more sophisticated theorizing about maritime security. Interdiction-based models are insufficient to meet demand, while competing incentives, interests, and political concerns define a context inimical to easy solutions. We saw yet again, moreover, the need for nuanced approaches that leverage military capacity to address issues at the "softer" end of the threat spectrum (what I have been calling quality-of-life issues) without eliciting concerns of over militarization.

This assessment of counterpiracy's real-world obstacles in the Straits of Malacca and Singapore is further reinforced in a survey of the literature on Southeast Asian maritime security, which is quite dynamic. This body of work, through an infusion of terrorism research into the maritime sphere, has come to organically echo elements of the broken windows theory laid out in chapter 2—multidimensionality and contextualization. It is increasingly clear that piracy, terrorism, narcotics, and more are not isolated issues (they are multidimensional) and that only by dismantling the infrastructure that undergirds this system can disorder be curtailed. Likewise, an appreciation for the utility of engaging with local communities and the importance of considering the manner in which enforcement is framed, marketed, and executed speak to the various ways that context and environment (both physical and psychological) should inform strategy. Discussing this literary evolution in the guise of the broken windows theory demonstrates how a more rigorous framing lends greater structure still to this dynamic body of work. Thus, from both real-world realities and academic theorizing, broken windows theory clearly provides a useful lens even in a region with long-standing investments and investigations into maritime security.

The *Ticonderoga*-class guided-missile cruiser USS *Princeton* (CG 59) transits the Singapore Strait. *Princeton* is part of the *Nimitz* carrier strike group on a regularly scheduled deployment in the U.S. Seventh Fleet area of responsibility in support of maritime security operations and theater security cooperation efforts. *U.S. Navy photo by Mass Communication Specialist 3rd Class Kelsey J. Hockenberger/Released*

CHAPTER EIGHT
Charting a Course

CRITICISM

The broken windows theory is not without its flaws, and neither is its legacy. In recent years, the theory has become almost a shorthand reference for biased policing and adversarial relationships between police and minority communities. These tensions burst to the fore in late 2014 after a series of unarmed black men were killed in police custody across the United States (including Eric Garner on Staten Island), bringing broken windows back into the public lexicon. Importantly, the invocation of broken windows as a reference to policies like New York's stop-and-frisk (which, I would argue, are anathema to the tenets of community policing, as explored in chapter 2) is not necessarily reflective of the theory or its early implementation. The vernacular and academic uses of broken windows have largely operated in separate spheres, but some notables in criminology, including the theory's cocreator, George Kelling, have advocated for a better distinction between failures in policing and the theory.[1]

Nevertheless, even if only in the pop lexicon, broken windows has been thrust into the social consciousness in a new way as the United States faces a growing movement to overhaul policing. And, in large part, police departments own this responsibility. Instead of elevating broken windows' emphasis on people and perceptions, some police departments invested deeper in zero-tolerance policies, taking their eyes off the sorts of issues communities care about most. In so doing, even as crime in cities like New York remained at record lows, police constructed an environment in which the community perceived the enforcement of law to be unequal. The very lessons in this book, which suggest an aversion to militarizing police and the harm of creating fortress mentalities, were lost on some over time. There are probably a lot of reasons for this—surplus military goods diverted to police forces after 9/11 certainly didn't help drive a demilitarized mentality in policing. Moreover, that citations, searches, and arrests have historically been skewed against minority communities may be more expressive of a broken system than any one broken theory.

Still, while these are important conversations, for our sake the debate over police in modern America is less relevant than the idea that we can learn something from criminology with applicability for the maritime services. In fact, that basic lesson is the most important.

In truth, though, I do hope you are sold that the broken windows theory provides a novel and useful conceptual lens for investigating maritime security. That lens enables us to come to two deceptively simple yet transformative conclusions—that crime is multidimensional and that environment matters. Or, as Kelling writes in his recent defense of the theory, "that small things matter in a community and, if nothing is done about them, they can lead to worse things."[2] It is this fundamental mixture that propelled the broken windows theory to popularity, especially in New York, and these are the core principles that I have applied to the littorals.

There are, to be sure, more specific criticisms of the ideas I've put forward in this book. Financial constraints made field research infeasible, a shortcoming I tried to correct by leveraging voices from within the communities under study. The very focus on theory creates an even more interesting dilemma of competing incentives: theory and strategy dictate that we find ways of explaining complicated phenomena while being as parsimonious and succinct as possible. Inevitably, the models we produce to do so miss things. In fact, they are designed to miss things. Models are supposed to help us cast aside everything but the factors that are essential for understanding a given environment—getting that right, however, is no easy task. Finally, reaching between disciplines invariably sparks controversy over how well some ideas export from one field to another, and it strains an author's ability to maintain currency across so many literatures. It is my hope that these critiques are mitigated, in part, by the depth of thoughtful commentary found in this book, the peer review to which it was subjected, and its research design in its previous incarnation as a doctoral thesis.

Summation

I opened this book with an illustration of a world far outside the minds of many American naval strategists but one well within the lived experience of millions of people around the world. Dense slums. Congested waterways.

Contested spaces. These conditions predominate in the lives of a rapidly growing percentage of the global population. Maritime security embodies the hybrid threat. And because such threats are so often manifested in criminality, those responsible for addressing insecurity in the littorals would, I suspected, be well served by an understanding of criminology. The broken windows theory is particularly fitting for this task. Just as it revolutionized a midcentury policing model built on fighting and interdicting crime, so too does broken windows offer a similarly intriguing redefinition of security in the littorals. It is this reality that inspired the kernel of my research question: how does this maritime space relate to the functions of navies today?

This core question grew in complexity and specificity until I arrived at the main point of inquiry, as laid out in the introduction: can the broken windows theory provide a lens through which to theorize maritime security in the littorals? To answer that question, however, requires a better understanding of two underlying issues. First, why should we care? What was so wrong with existing theories, or so different about the world today, that makes an investigation like this relevant? Second, what is the broken windows theory? Without understanding the fundamentals of this theory, we would be unable to apply its lens in the realm of maritime security.

First, what in the strategic literature made this research relevant to begin with? I address that in two sections in chapter 1. First, I elucidated the threat forecast from which we are operating. This forecast illustrates the major trends shaping the future of the littoral space over the coming decades, including Kilcullen's four megatrends: population growth, urbanization, littoralization, and connectedness. This analysis not only tells us where to expect future tensions to arise (in the littorals) but hints at the cause of these hot spots (transnational issues of human insecurity). It is in this conversation that we were introduced to the idea of feral cities, urbicide, and hybrid threats, which we revisited over the course of the book.

In the second portion of chapter 1, I turned toward an analysis of some strategic literature to pull out some trends in the emergence of maritime security as a construct. We saw how strategies increasingly address maritime security in broad strokes but leave unanswered fundamental questions about what maritime security means. Strategic documents from NATO, the European Union, and the African Union similarly offer little guidance on some of the larger questions of theory and strategy—maritime security remains a bucket of things that seem important but don't really fit anywhere else. Those strategies that do devote greater space to maritime security most often read like operational laundry lists and are consistently framed around specific threat vectors (e.g., counterpiracy or counternarcotics) without bridging between them. Chapter 1 ends with a more detailed look at one particular strategy, *A Cooperative Strategy for 21st Century Seapower* (*CS21*), which was in some ways a watershed moment for maritime security. Yet, even in that exploration we see a fundamental, expected, and probably warranted but no less important dichotomy between the strategy's language on maritime strategy, on the one hand, and acquisitions priorities, on the other. After all that exploration, what emerges is a strategic space without much theory, a strategic headline to which writers widely refer but which lacks any deeper coherence. To me, that sounds like reason enough to start talking about maritime security strategy.

In chapter 2, I began with an exploration of the theory at large, dissecting some of the central tenets of broken windows. To offer a deeper familiarity with the theory and its literature, I turned to an illustration of its use in the New York City subway. In so doing, we not only explored a common case study in broken windows scholarship but also saw the theory depicted as a natural experiment. This is the first major step in demonstrating the exportability of a criminological theory beyond the urban space and into a maritime one. That extrapolation was then reinforced in the discussion of broken windows in the greater New York City environment, which gave us a chance to further explore the mechanisms of broken windows policing. At that point, we were already starting to see the skeleton of the theory take shape, so to help us get to the essence of the theory, I turned the conversation toward the use of broken windows across other disciplines. As a result, we could see with certainty that the broken windows theory is not specific to cities or crime. We could see that the theory speaks to broader psychological and sociological states—ones we could apply to something like the littorals. Whether fighting crime in New York, venereal diseases on the streets of Louisiana, or pirates off the Gulf of Guinea, broken windows' central hypotheses are useful lenses for understanding the actions and conditions of the world around us.

Across the discussion in chapter 2, I keyed in on two hypotheses, which I carried throughout this book—that crime is multidimensional and that crime is context dependent. In the subway, in New York City, and among seminary students and middle schoolers, these two elements play an ever-present role in our understanding of disorder. And while debate remains surrounding the particular pathways of cause and influence of these hypotheses in the criminological literature, of those two elements we could be certain. So I closed chapter 2 by acknowledging the ongoing conversation within the theory while underscoring the validity of adopting broken windows' central tenets as a foundation for understanding insecurity globally.

Having addressed the underlying questions of why and what, we turned in chapter 3 to the heart of that question. How can the broken windows theory provide a lens through which to theorize maritime security in the littorals? In response, the Caribbean Basin case explores in granular detail the life, crimes, and transnational flows of actors across the region in an effort to highlight the two broken windows themes in chapter 2: that context influences action and that such actions are not isolated. Through a look at narcotics trafficking, local violence, irregular migration, terrorism, corruption, and more, I repeatedly illustrated in part 2 of this book the applicability and utility of this lens. Maritime insecurity across the Basin is shown to be heavily multidimensional, with social cues shaping not only crime in the Caribbean but also recommendations for communal enforcement models. Local murders interplay with the transnational flow of drugs. Crimes committed within a community compound to create a chain of criminality extending across continents. People moving irregularly in search of better lives obscure deeper illicit trafficking flows. The transnational distribution of money throughout the Caribbean lubricates crime while corroding faith in political institutions. Terrorists are allegedly granted access to fundraising and resources through the same infrastructure maintained by criminal organizations. Local contexts—crime, poverty, corruption—contribute to how Caribbean authorities conceptualize security, while multidimensional actors frustrate enforcement efforts centered on interdiction or single threats.

The relevance of this broken windows approach is reinforced by our exploration of local perspectives on transnational crimes, which illustrates that many in the Basin have long viewed security as multidimensional. Caribbean law enforcement has already embraced community policing's precepts in region-wide trainings and seminars, validating further the utility of appropriating the broken windows theory as a lens for thinking about maritime disorder. The role of perception also permeates this conversation, providing a new lexicon for discussing littoral security and allowing us to identify subtle themes across the literature. In so doing, we could see the environment as a domain in and of itself—a contested space of ideas, behaviors, and physical signals that inform the actions of those living within the Basin. And so, with part 2 of this book, we have a proof of concept, demonstrating that we could indeed export the broken windows theory's ideas to the maritime space.

By the end of the second third of this book, we had explored what the theory is, why it matters, and that it does provide an interesting perspective through which to evaluate maritime security in the Caribbean. What remained was some sort of evidence that these findings had value in the wider world, to generalize this concept beyond one case. So together we embarked on the final portion of this book, into the first of two illustrative cases—chapter 6, the Gulf of Guinea. In this study, I leaned on the criminal relationships and vocabulary established in the Caribbean study to identify commonalities between the two regions. By illustrating the similarities between Caribbean and West African trade in narcotics as well as the myriad quality-of-life crimes endemic to both regions, we built strong circumstantial evidence that conclusions from the former case could be applied to the latter.

Yet this was only half of the effort. In chapter 6 I also introduced one variant of maritime disorder that we had left previously untouched—piracy. A new problem is a wonderful test for a theory. If we could fold that problem into the existing theory, this would suggest that the strategic framework I was trying to build was in good shape. If not—well, this book would have had a different ending. By illustrating that piracy can be integrated into the same narrative on maritime crime's multidimensional and contextual nature, we see that a broken windows lens is capable of adapting to new issues in new places and thus is not unique in its application to the Caribbean. In other words, good news on the generalizability front.

Still, two examples don't make a trend, so we turned to another illustrative case study to offer a slightly tweaked version from that prior. Around the Straits of Malacca and Singapore, piracy is an issue of relatively long-standing study and state action. Moreover, in this region piracy was even successfully suppressed for a time, or at least piracy numbers declined. The challenge here was to see if broken windows has anything useful to contribute to a region that has been thinking about maritime security, particularly piracy, for quite a while. And, indeed, it does. As it turns out, the phenomenon of piracy in Southeast Asia lends itself well to an analysis through a broken windows lens. Further still, when reading the academic literature on the region through a broken windows perspective, we can see very clear but unstated currents among analysts who are increasingly talking about multidimensional and environmental themes. Thus, even in a region long accustomed to discussing and responding to piracy, a broken windows lens provides a new and interesting perspective.

FINDINGS AND CONSEQUENCES

But so what? Ultimately, a theoretical lens, however interesting, has to provide some actionable results. To better parse this, it is useful to think of this book as speaking to two audiences. First, for the academic, the multidisciplinary nature of this book serves a bridging role for a variety of streams of literature—from security studies to naval research to criminology. While it is impossible to address sufficiently every debate upon which this research touches, I endeavored throughout to highlight linkages across disciplines. Second, the topics covered here have clear overtones for naval strategists, warfighters, and policymakers. As I discuss at length in chapter 1, littoral security will likely play a dominant role in the twenty-first century, regardless of some institutional disinterest in the gray spaces of order maintenance. For both audiences, this research fills a void in the emerging field of maritime security, which is noticeably absent of strategic frameworks. It is my hope that such a framework will prove useful for both communities. To help in that process, it would be instructive to identify some of the most significant conclusions this research has for each audience.

For Researchers
Briefly, it is instructive to reiterate many of the debates upon which this research touched and informs. To begin, this book helped bring the study of maritime security into greater contact with an expanding literature on transnational threats. Chapter 7 serves as an archetypical example of the evolving role transnational issues play in the way we understand maritime insecurity. There I detailed how a focus on piracy in Southeast Asia merged after 9/11 with an interest in terrorism and eventually a wide variety of nonstate threats, yielding over time a multidimensional understanding of the complex world of illicit international flows. Since 9/11, there has been a concerted effort by some to bring the transnational world into greater relief. And yet this body of work often remains isolated from wider security studies literature, particularly maritime security, which is itself already an outlier in many security studies programs. However, the point, made in practice throughout this book, is that any one portion of security or maritime research is incomplete without an accompanying transnational component. It is among my greatest hopes that this research helps facilitate an acceleration of transnational maritime security considerations in the academic and strategic consciousness.

Of course, any consideration of nonstate maritime threats leads implicitly toward a conversation on the role of navies in peacetime, a theme found throughout this book and yet another ongoing debate in policy and academic circles. Beginning in chapter 1, I suggest that the very concept of peacetime versus wartime operations is an outdated way of understanding nonstate conflict. From the "savage wars of peace" to the demands of stability and development operations, I underscore at the outset that the conventional paradigm of war and peace loses some relevance in a transnational framework. Significantly, there was a brief moment of triumph for this framing in the maritime service's 2007 capstone document (*CS21*). In the context of maritime security, the maritime services hint in *CS21* at a drift away from the war and peace dichotomy toward an appreciation of order maintenance and conflict prevention. Still, as I conclude, the Navy has yet to metaphorically "put its money where its mouth is" when tackling maritime security, with respect to both monetary and intellectual investments, and renewed emphasis on preparing for major power competition will not work in maritime security's favor. Still, the debate over ambiguous security postures (peace versus war) continues to surface across this book. In chapter 2, the role of police in keeping the peace is similar in essential character to the military debate over a less binary perspective on war and peace. In parts 2 and 3 I also touched on themes—such as urbicide and the conflicting demands of using military equipment and personnel to counter criminal actors—with obvious implications for the debate on peacetime military (and policing) operations. For nearly two decades the United States has struggled to understand how a seemingly endless war on terrorism fits—legally, ethically, and strategically—into a world order built around a black-and-white understanding of conflict. This book does not resolve that debate, but it is clearly contoured by it.

Finally, in a narrower maritime lens, the Navy's peacetime responsibilities are equally relevant for an area in which there should be significantly more debate—what does good order at sea mean, strategically? Some, as noted in chapter 2, regard the phrase "maritime security" largely as a rebranding of the more conventional "good order at sea." Yet, no matter the language we use, it remains clear that promoting good order is a particularly sparse arena of strategic debate in the naval world. It makes sense that a theory of order maintenance (e.g., the broken windows theory) would provide a foundation upon which to lend greater conceptual depth to such a long-standing but poorly framed notion. In that guise, this book serves as a potential launching point for further theoretical work on good order at sea. Perhaps in so doing, we can even come to a better understanding of just how "good order at sea" and "maritime security" actually relate to one another.

For Strategists

Understanding, now, how the broken windows theory helps navies think about littoral spaces and the insecurities that inhabit them—what does this mean for U.S. strategy in practice? To answer that question, we need to revisit the fundamentals of broken windows as expressed in chapter 2. Community policing is by nature a problem-oriented approach. Its central tenets are therefore not as overtly prescriptive as perhaps Mahanian precepts to control sea and trade. Instead, in every city, in every precinct, even on every block, broken windows armed police with the means to identify problems in conjunction with community members and to understand how enforcement decisions shape communal self-efficacy. Thus, I have so far avoided making too many specific operational recommendations. Still, while the findings I reach here are not as prescriptive as other strategic frameworks may be, they are nevertheless consequential for how the U.S. Navy should perceive its role operating in crowded littorals. I conclude, therefore, with some practical strategic guidance.

First, navies need to understand and adapt to the dichotomous demands of military objectives, on one hand, and law enforcement (or peacetime) objectives, on the other. While pursuing the latter may require military technical capabilities, given the advancing sophistication of non-state threats, these two objectives require patently different ways of interacting with communities (think of the consequences of the Tivoli Gardens raid in Jamaica, for example). The U.S. Navy should continue to invest in the idea of diverging strategies to meet these divergent objectives, embracing the groundwork laid by *CS21*'s implicit lean toward two complementary but distinct strategies—one for regional engagement, one for global competition.

Second, the U.S. Navy cooperates extensively with local authorities across the world. A broken windows–inspired approach would require, however, that policymakers further elevate local partners as the focal point of littoral strategies. Just as police, in community-oriented models, let neighborhoods help set the enforcement agenda to meet their fears and concerns, so too does a navy operating in littoral spaces need to work in line with local social objectives. In many cases those local (and national) objectives vary significantly from American strategic considerations. Consider, for example, the diverging security priorities even of the three littoral states in the Strait of Malacca case, detailed in chapter 7. This should consequently inform the third consideration—remember multidimensionality.

Even if communities or nations have different first-order concerns than those of the United States' maritime forces, the multidimensional nature of littoral crime sketched throughout this book reinforces the notion that nothing is in fact isolated. Combating illegal fishing may precipitate declines in piracy, and combating human trafficking may reduce the spread of public health concerns. The point for strategists is that focusing on placing local communities in the driver's seat and building communal self-efficacy—no matter the individual points of concern—helps breed greater stability across a spectrum of issues. In other words, navies and coast guards should emphasize the positive mobilization of maritime communities in pursuit of their own security. Partnership is not zero-sum, and pursuing a local agenda helps ensure long-term buy-in from local authorities and the people they will rely on to help keep communities safe.

Finally, do not neglect context and perceptions. Missions on the softer end of the operational spectrum "are not ancillary to the broader strategic landscape, but pivotal in its construction."[3] Unlike in blue waters, maritime forces in the littorals will routinely find themselves operating with and around maritime communities. Without a clear understanding of how enforcement at the communal level shapes broader disorder, these forces will inevitably fail to achieve long-term strategic success. In planning deployments in the littorals, the U.S. Navy should prioritize understanding and positively impacting local perceptions. This is, ultimately, at the root of any community policing objective—the endeavor to help communities feel as if they have a stake in their neighborhoods. That objective is as consequential to maritime security as it is to urban crime control.

Chris Trelawny, a Royal Navy reservist and International Maritime Organization staffer, writes that he senses "an almost existential crisis around the question of what navies are for," with grand strategies often struggling under the weight of "governance 'by accountant.'"[4] And while, as I laid out in chapter 1, the central role of the U.S. Navy and some of its smaller cousins is and will remain to deter and (if need be) win large-scale conflicts, the twenty-first century presents a threat forecast replete with risks that do not conform to convention. As we have explored, there is a growing body of work that underscores that capable navies would be well served to help states in troubled regions develop constabulary functions to address the instability that disorder can breed.[5] This book is an attempt to jumpstart the conversation on what that means, strategically, and how the U.S. Navy and its sister forces might start thinking about these issues outside the tactical and operational bubble. The tides are shifting, and the littorals are reemerging. Conceptualizing muddy waters and the opaque world of crime and war they harbor is paramount.

Notes

Introduction
1. "Slum Almanac 2015–2016," United Nations Habitat web page, https://unhabitat.org/slum-almanac-2015-2016/.
2. The details for this reconstruction come primarily from David Kilcullen, *Out of the Mountains: The Coming Age of the Urban Guerrilla* (London: Hurst, 2013), 54–65. While Kilcullen's telling is particularly insightful for its emphasis on the littoral component, the 2008 Mumbai attack is the source of a lot of scholarship. With respect to feral cities, which I explore in chapter 2, see John Sullivan and Adam Elkus, "Postcard from Mumbai: Modern Urban Siege," *Small Wars Journal*, February 16, 2009. With respect to an operational assessment, see Sarita Azad and Arvind Gupta, "A Qualitative Assessment on 26/11 Mumbai Attack Using Social Network Analysis," *Journal of Terrorism Research* 2, no. 2 (2011), https://cvir.st-andrews.ac.uk/articles/10.15664/jtr.187/. With respect to the technical component of the attack, see Marc Goodman, *Future Crimes: Everything Is Connected, Everyone Is Vulnerable, and What We Can Do about It* (New York: Doubleday, 2015), 81–84.
3. Kilcullen, *Out of the Mountains*, 65.
4. Kilcullen, 65.
5. Kilcullen, 54–55.
6. Quoted in Geoffrey Till, *Seapower: A Guide for the Twenty-First Century*, 3rd ed. (New York: Routledge, 2013), 346.
7. Quoted in Till, 346.
8. Till, 83. See the British doctrinal definition: Till, 267.
9. Kilcullen, *Out of the Mountains*, 30.
10. Till, *Seapower*, 343.
11. Till, 257.
12. Kilcullen, *Out of the Mountains*, 269.
13. Christian Bueger, "What Is Maritime Security?," *Marine Policy* 53 (2015): 160; and Basil Germond, "The Geopolitical Dimension of Maritime Security," *Marine Policy* 54 (2015): 137.

14. Till, *Seapower*, 25, 283.
15. Till, 24.
16. Germond, "The Geopolitical Dimension of Maritime Security," 138. Maritime security has also been defined in the negative, as a lack of threats or insecurity. Freedom Onuoha, for example, of Nigeria's National Defense College, defines it as "the freedom from or absence of those acts that could negatively impact the natural integrity and resilience of any navigable waterway or undermine the safety of persons, infrastructure, cargo, vessels, and other conveyances." See Freedom Onuoha, "Piracy and Maritime Security off the Horn of Africa: Connections, Causes, and Concerns," *African Security* 3, no. 4 (2010): 193.
17. Bueger, "What Is Maritime Security?," 161.

Chapter One. Shifting Tides

1. Department of Defense, *Quadrennial Defense Review* 2014 (Washington, D.C.: Department of Defense, 2014), http://archive.defense.gov/pubs/2014_Quadrennial_Defense_Review.pdf, III.
2. David Kilcullen, *Out of the Mountains: The Coming Age of the Urban Guerrilla* (London: Hurst, 2013), 239.
3. Kilcullen, 25.
4. Kilcullen, 29.
5. Mike Davis, *Planet of Slums* (New York: Verso, 2007), 20.
6. Kilcullen, *Out of the Mountains*, 29.
7. Dr. Thomas Mahnken (former deputy assistant secretary of defense for policy planning), discussion with the author, June 24, 2014.
8. Kilcullen, *Out of the Mountains*, 247.
9. Richard Norton, "Feral Cities," *Naval War College Review* 56, no. 4 (2003): 98.
10. John Sullivan and Adam Elkus, "Postcard from Mumbai: Modern Urban Siege," *Small Wars Journal*, February 16, 2009, 8.
11. Matthew Frick, "Feral Cities—Pirate Havens," U.S. Naval Institute *Proceedings* 134, no. 12 (2008), http://www.usni.org/magazines/proceedings/2008-12/feral-cities-pirate-havens; and Ken Stier, "Feral Cities," *New York Times*, December 12, 2004, http://www.nytimes.com/2004/12/12/magazine/12FERAL.html?_r=0.
12. Kilcullen, *Out of the Mountains*, 40.
13. Kilcullen, 114.
14. Kilcullen, 89, 93.
15. Kilcullen, 141.
16. Kilcullen, 45.
17. Kilcullen, 99.
18. Sullivan and Elkus, "Postcard from Mumbai."

19. Frank Hoffman, "Hybrid vs. Compound War," *Armed Forces Journal*, October 1, 2009, http://www.armedforcesjournal.com/hybrid-vs-compound-war/.
20. Hoffman.
21. Barack Obama, *National Security Strategy* (Washington, D.C.: White House, May 2010), https://obamawhitehouse.archives.gov/sites/default/files/rss_viewer/national_security_strategy.pdf, 49.
22. Kilcullen, *Out of the Mountains*, 126.
23. Kilcullen, 287.
24. Kilcullen, 287. Dr. Mahnken, in conversation with the author, likewise identifies the role Italian and Australian police play in augmenting American law enforcement capabilities abroad. Mahnken offers that the lack of an American federal police force may hinder the United States' ability to train and advise law enforcement overseas and creates a capabilities gap often filled by the Italians or Australians.
25. Sullivan and Elkus, "Postcard from Mumbai," 12.
26. Kilcullen, *Out of the Mountains*, 287.
27. Kilcullen, 23.
28. Kilcullen, 24.
29. Kilcullen, 24.
30. Kilcullen, 23; and John Shy, "The American Military Experience: History and Learning," *Journal of Interdisciplinary History* 1, no. 2 (1971): 213.
31. Kilcullen, *Out of the Mountains*, 23.
32. Kilcullen, 24.
33. Kilcullen, 261.
34. Sullivan and Elkus, "Postcard from Mumbai," 6.
35. Eyal Weizman, "The Art of War," *Frieze* 99 (May 2006), https://frieze.com/article/art-war.
36. Sullivan and Elkus, "Postcard from Mumbai," 7–8. Emphasis in original.
37. Kilcullen, *Out of the Mountains*, 19–20.
38. Kilcullen, 111.
39. Kilcullen, 27.
40. Shy, "The American Military Experience," 223.
41. Shy, 224.
42. Shy, 223.
43. John Nagl, *Learning to Eat Soup with a Knife: Counterinsurgency Lessons from Malaya and Vietnam* (Chicago: University of Chicago Press, 2002), 223.
44. Alfred Thayer Mahan, *The Influence of Sea Power upon History: 1660–1783* (New York: Dover, 1987), 288.
45. Nagl, *Learning to Eat Soup with a Knife*, 49.
46. Geoffrey Till, *Seapower: A Guide for the Twenty-First Century*, 3rd ed. (New York: Routledge, 2013), 48.

47. Till, 48.
48. Till, 57.
49. Till, 61.
50. Till, 62.
51. Hugh White, "Navy on the Wrong Course Building Large Warships," *Sidney Morning Herald* (Australia), October 29, 2013, https://www.smh.com.au/opinion/navy-on-the-wrong-course-building-large-warships-20131028-2wbu0.html.
52. Till, *Seapower*, 35.
53. Till, 36.
54. Quoted in Till, 36.
55. Till, 36.
56. Till, 348.
57. Kilcullen, *Out of the Mountains*, 16–17.
58. Kilcullen, 16–17.
59. Kilcullen, 114.
60. White House, *National Strategy for Maritime Security* (Washington, D.C.: White House, September 2005), https://www.state.gov/documents/organization/255380.pdf, 7–8.
61. In 2015 the maritime services published a revised cooperative strategy, to which the author responded in a post on CIMSEC's *NextWar* blog. See Joshua Tallis, "The New US Maritime Strategy," Center for International Maritime Security (CIMSEC) *NextWar* (blog), March 13, 2015, http://cimsec.org/new-us-maritime-strategy/15507. However, for the purposes of understanding the impact of *CS21* on acquisitions and strategic guidance, we focus on the original document since the revision had not existed long enough to potentially direct policy at the time of writing.
62. Till, *Seapower*, 51.
63. Basil Germond, "The Geopolitical Dimension of Maritime Security," *Marine Policy* 54 (2015), 137.
64. On maritime drug smuggling, see J. Ashley Roach, "Initiatives to Enhance Maritime Security at Sea," *Marine Policy* 28 (2004): 45.
65. Roach, 46.
66. Kilcullen, *Out of the Mountains*, 279.
67. Kilcullen, 261.
68. Department of the Navy, . . . *From the Sea: Preparing the Naval Service for the 21st Century* (Washington, D.C.: Navy News Service, September 1992), http://www.navy.mil/navydata/policy/fromsea/fromsea.txt. See also Department of the Navy, *Forward . . . From the Sea: The Navy Operational Concept* (Washington, D.C.: Department of the Navy, 1997).
69. U.S. Marine Corps, *Marine Corps Operations*, MCDP 1-0 (Washington, D.C.: HQ Marine Corps, 2011), 1-15, 2-20.

70. Till, *Seapower*, 274.
71. Till, 84.
72. Department of Defense, *A Cooperative Strategy for 21st Century Seapower* (Washington, D.C.: Department of Defense, October 2007).
73. Robert Work, *Naval Transformation and the Littoral Combat Ship* (Washington, D.C.: Center for Strategic and Budgetary Assessments, February 2004), 131.
74. Work, 2.
75. Loren Thompson, "Navy Has Few Options if Littoral Combat Ship Falters," *Forbes*, March 7, 2014, http://www.forbes.com/sites/lorenthompson/2014/03/07/navy-has-few-options-if-littoral-combat-ship-falters/.
76. John Kirby, "The Littoral Combat Ship: Give It Time," *Information Dissemination* (blog), June 10, 2013, http://www.informationdissemination.net/2013/06/the-littoral-combat-ship-give-it-time.html.
77. Thompson, "Navy Has Few Options."
78. Till, *Seapower*, 123.
79. Thompson, "Navy Has Few Options."
80. Peter Haynes, *Toward A New Maritime Strategy: American Naval Thinking in the Post–Cold War Era* (Annapolis, MD: Naval Institute Press, 2015), 236–37.
81. Department of Defense, *Quadrennial Defense Review 2014*, 22, 59.
82. Department of Defense, *A Cooperative Strategy for 21st Century Seapower*.
83. Obama, *National Security Strategy*, 1.

Chapter Two. Breaking Windows

1. Malcolm Gladwell, *The Tipping Point: How Little Things Can Make a Big Difference* (Boston: Back Bay Books, 2002), 19.
2. Gladwell, 158.
3. Gladwell, 165.
4. William Sousa and George Kelling, "Of 'Broken Windows,' Criminology, and Criminal Justice," in *Police Innovation: Contrasting Perspectives*, ed. David Weisburd and Anthony Braga (Cambridge: Cambridge University Press, 2006), 77.
5. In fact, the view of police we are familiar with today, the crime-fighting model, is actually not traditional. It is the result of a reformist movement in American politics, which finds its roots in the Progressives of the early twentieth century and reaches fruition in midcentury. Before then police were charged with maintaining order, much in the way broken windows advocates. For a history, see George Kelling and Catherine Coles, *Fixing Broken Windows: Restoring Order and Reducing Crime in Our Communities* (New York: Touchstone, 1996), chap. 3; and for more resources, see Wesley Skogan, *Disorder and Decline: Crime and the Spiral of Decay in*

American Neighborhoods (Berkeley: University of California Press, 1990), 85–90.
6. James Wilson and George Kelling, "Broken Windows: The Police and Neighborhood Safety," *Atlantic*, March 1, 1982, http://www.theatlantic.com/magazine/archive/1982/03/broken-windows/304465/.
7. Wilson and Kelling, 2.
8. Wilson and Kelling, 2.
9. Wilson and Kelling, 2. Skogan likens disorder to a disease, calling petty crimes such as vandalism and littering "contagious" to explain the tendency for disorder to beget evermore disorder. See Skogan, *Disorder and Decline*, 49.
10. Wilson and Kelling, "Broken Windows," 2.
11. Kelling and Coles, *Fixing Broken Windows*, 98.
12. Kelling and Coles, 70.
13. Kelling and Coles, 144.
14. George Kelling and William Bratton, "Declining Crime Rates: Insiders' View of the New York City Story," *Journal of Criminal Law and Criminology*, no. 4 (1998): 1219.
15. George Kelling and William Sousa Jr., *Do Police Matter? An Analysis of the Impact of New York City's Police Reforms*, Civic Report no. 22 (New York: Manhattan Institute, 2001), Executive Summary.
16. John Shy, "The American Military Experience: History and Learning," *Journal of Interdisciplinary History* 1, no. 2 (1971): 210.
17. Kelling and Coles, *Fixing Broken Windows*, 71.
18. Kelling and Coles, 83.
19. Kelling and Coles, 38.
20. In *Fixing Broken Windows* there are more than twenty individual cases cited throughout the text. For a list of all cited cases, see the Civil Liberties/Judicial System entry on page 305 of the index.
21. Kelling and Coles, 162.
22. Gladwell, *Tipping Point*, 136.
23. Gladwell, 136.
24. Kelling and Coles, *Fixing Broken Windows*, 118; and Gladwell, *Tipping Point*, 137.
25. Gladwell, *Tipping Point*, 137.
26. Wilson and Kelling, "Broken Windows," 2.
27. Kelling and Coles, *Fixing Broken Windows*, 116; and Gladwell, *Tipping Point*, 143.
28. Kelling and Coles, *Fixing Broken Windows*, 117; and Kelling and Bratton, "Declining Crime Rates," 1221.
29. Kelling and Bratton, "Declining Crime Rates," 1221.
30. Kelling and Coles, *Fixing Broken Windows*, 121.

31. Kelling and Coles, 125, 131.
32. Gladwell, *Tipping Point*, 143; and Kelling and Coles, *Fixing Broken Windows*, 132.
33. Kelling and Bratton, "Declining Crime Rates," 1221; and Gladwell, *Tipping Point*, 144.
34. Gladwell, *Tipping Point*, 144–45.
35. Kelling and Coles, *Fixing Broken Windows*, 134.
36. Kelling and Coles, 137.
37. Gladwell, *Tipping Point*, 145.
38. Kelling and Bratton, "Declining Crime Rates," 1222.
39. Kelling and Coles, *Fixing Broken Windows*, 137.
40. Kelling and Bratton, "Declining Crime Rates," 1222–24.
41. Robert Sampson and Stephen Raudenbush, "Systematic Social Observation of Public Spaces: A New Look at Disorder in Urban Neighborhoods," *American Journal of Sociology*, no. 3 (1999): 604.
42. Kelling and Coles, *Fixing Broken Windows*, 141–42.
43. Kelling and Coles, 142–43.
44. Gladwell, *Tipping Point*, 146.
45. Kelling and Bratton, "Declining Crime Rates," 1217–18.
46. Gladwell, *Tipping Point*, 5–6.
47. Kelling and Coles, *Fixing Broken Windows*, 153.
48. Kelling and Coles, 146.
49. Kelling and Coles, 147.
50. Kelling and Bratton, "Declining Crime Rates," 1228; and Kelling and Sousa, *Do Police Matter*, 10.
51. Kelling and Sousa, *Do Police Matter*, 10.
52. David Kilcullen, *Out of the Mountains: The Coming Age of the Urban Guerrilla* (London: Hurst, 2013), 288.
53. Kelling and Bratton, "Declining Crime Rates," 1226.
54. Kelling and Coles, *Fixing Broken Windows*, 144. Skogan also lists decentralization as one of his four principles of community policing. The others include commitment to problem-oriented policing, addressing community demands, and helping the community become a partner in its own defense. See Skogan, *Disorder and Decline*, 91–92.
55. Kelling and Coles, *Fixing Broken Windows*, 159, 170.
56. Kelling and Sousa, *Do Police Matter*, 16–17.
57. John Nagl, *Learning to Eat Soup with a Knife: Counterinsurgency Lessons from Malaya and Vietnam* (Chicago: University of Chicago Press, 2002), 74.
58. Quoted in Nagl, 156.
59. Nagl, 43.
60. Daniel Wilner, Rosabelle Walkley, John Schram, Thomas Pinkerton, and Matthew Tayback, "Housing as an Environmental Factor in Mental Health:

The Johns Hopkins Longitudinal Study," *American Journal of Public Health and the Nation's Health* 50, no. 1 (1960): 60.
61. Deborah Cohen, Suzanne Spear, Richard Scribner, Patty Kissinger, Karen Mason, and John Wildgen, "'Broken Windows' and the Risk of Gonorrhea," *American Journal of Public Health*, no. 2 (2000): 230–31.
62. The broken windows index was a scaling of multiple physical neighborhood characteristics. It was tabulated in part by referring to videos made of every street in the fifty-five block groups, assigning a number ranging from one to four to assess no or mild cosmetic damage, to minor or major structural damage. That information was combined with coded reports on public school conditions made by the city's sanitation department, as well as assessments created by walking the streets to collect data on trash and graffiti. Cohen et al., 231.
63. Cohen et al., 232.
64. Tod Mijanovich and Beth Weitzman, "Which 'Broken Windows' Matter? School, Neighborhood, and Family Characteristics Associated with Youths' Feelings of Unsafety," *Journal of Urban Health* 80, no. 3 (2003): 401.
65. Mijanovich and Weitzman, 413.
66. Mijanovich and Weitzman, 403.
67. Skogan, *Disorder and Decline*, 47, 74.
68. Sampson and Raudenbush, "Systematic Social Observation of Public Spaces," 608.
69. Sampson and Raudenbush, 603, 608, 622.
70. Sampson and Raudenbush, 608.
71. Sampson and Raudenbush, 626.
72. Sung Joon Jang and Byron Johnson, "Neighborhood Disorder, Individual Religiosity, and Adolescent Use of Illicit Drugs: A Test of Multilevel Hypotheses," *Criminology*, no. 1 (2001): 114.
73. Jang and Johnson, 114.
74. Sousa and Kelling, "Of 'Broken Windows,'" 84–85.
75. Skogan, *Disorder and Decline*, 73–75.
76. Howard Frumkin, "'Broken Windows': Frumkin Responds," *Environmental Health Perspectives* 113, no. 10 (2005): A657; and Cohen et al., "'Broken Windows' and the Risk of Gonorrhea," 235.
77. Nate Silver, *The Signal and the Noise: The Art and Science of Prediction* (New York: Penguin, 2013), 163–65; and Skogan, *Disorder and Decline*, 9.
78. Bernard Harcourt and Jens Ludwig, "Broken Windows: New Evidence from New York City and a Five-City Social Experiment," *University of Chicago Law Review*, no. 1 (2006): 298.
79. Silver, *The Signal and the Noise*, 185–86.

80. Leo Meléndez, "Disease and 'Broken Windows,'" *Environmental Health Perspectives* 113, no. 10 (2005): A657.
81. Gladwell, *Tipping Point*, 162.
82. Preeti Chauhan, Magdalena Cerdá, Steven Messner, Melissa Tracy, Kenneth Tardiff, and Sandro Galea, "Race/Ethnic-Specific Homicide Rates in New York City: Evaluating the Impact of Broken Windows Policing and Crack Cocaine Markets," *Homicide Studies* 15, no. 3 (2011): 268.
83. Chauhan et al., 279.
84. Gladwell, *Tipping Point*, 140.
85. Gladwell, 6.
86. Gladwell, 140.
87. Gladwell, 6.
88. Kelling and Sousa, *Do Police Matter*, 19.
89. Wilner et al., "Housing as an Environmental Factor in Mental Health."
90. Adam Crawford and Steven Hutchinson, "Mapping the Contours of 'Everyday Security': Time, Space and Emotion," *British Journal of Criminology*, December 12, 2015, 2, https://doi.org/10.1093/bjc/azv121.
91. Crawford and Hutchinson, 2.
92. Ben Brown, "Community Policing in Post-September 11 America: A Comment on the Concept of Community-Oriented Counterterrorism," *Police Practice and Research* 8, no. 3 (2007): 239–51.
93. Marie Breen-Smyth, "Theorising the 'Suspect Community': Counterterrorism, Security Practices and the Public Imagination," *Critical Studies on Terrorism* 7, no. 2 (2013): 223–40.
94. Bryn Caless, "Book Review: Martin Innes (2014). *Signal Crimes: Social Reactions to Crime, Disorder and Control*," *Policing*, December 2014, 1.
95. Martin Innes and Nigel Fielding, "From Community to Communicative Policing: 'Signal Crimes' and the Problem of Public Reassurance," *Sociological Research Online* 7, no. 2 (August 2002), http://www.socresonline.org.uk/7/2/innes.html.
96. Geoffrey Till, *Seapower: A Guide for the Twenty-First Century*, 3rd ed. (New York: Routledge, 2013), 25.
97. Pedro Cabral, Gabriela Augusto, Mussie Tewolde, and Yikalo Araya, "Entropy in Urban Systems," *Entropy*, no. 15 (2013): 5224, https://doi.org/10.3390/e15125223.
98. Kilcullen, *Out of the Mountains*, 42.

Chapter Three. The Business of Drugs
1. Ivelaw Griffith, "Probing Security Challenge and Change in the Caribbean," in *Caribbean Security in the Age of Terror: Challenge and Change*, ed. Ivelaw Griffith (Kingston, Jamaica: Ian Randle, 2004), 9.

2. Griffith, 4.
3. Griffith, 31; Trevor Munroe, "The Menace of Drugs," in *Caribbean Security in the Age of Terror: Challenge and Change*, ed. Ivelaw Griffith (Kingston, Jamaica: Ian Randle, 2004), 171; Norman Girvan, "Agenda Setting and Regionalism in the Greater Caribbean: Responses to 9/11," in *Caribbean Security in the Age of Terror: Challenge and Change*, ed. Ivelaw Griffith (Kingston, Jamaica: Ian Randle, 2004), 329; Isabel Jaramillo Edwards, "Coping with 9/11: State and Civil Society Responses," in *Caribbean Security in the Age of Terror: Challenge and Change*, ed. Ivelaw Griffith (Kingston, Jamaica: Ian Randle, 2004), 377; and Ivelaw Griffith, "Conclusion: Contending with Challenge, Coping with Change," in *Caribbean Security in the Age of Terror: Challenge and Change*, ed. Ivelaw Griffith (Kingston, Jamaica: Ian Randle, 2004), 513.
4. Girvan, "Agenda Setting and Regionalism in the Greater Caribbean," 329.
5. Griffith, "Probing Security Challenge," 10–11.
6. Clifford Griffin, "Regional Law Enforcement Strategies in the Caribbean," in *Caribbean Security in the Age of Terror: Challenge and Change*, ed. Ivelaw Griffith (Kingston, Jamaica: Ian Randle, 2004), 487.
7. Munroe, "The Menace of Drugs," 169.
8. Munroe, 168.
9. Munroe, 166; and State Department, *Patterns of Global Terrorism 2002* (Washington, D.C.: State Department, April 2003), https://www.state.gov/j/ct/rls/crt/2002/html/index.htm.
10. Munroe, "The Menace of Drugs," 170.
11. Griffith, "Probing Security Challenge," 27.
12. Griffith, 30–31.
13. Griffith, 31.
14. Richard Millett, "Weak States and Porous Borders: Smuggling along the Andean Ridge," in *Transnational Threats: Smuggling and Trafficking in Arms, Drugs, and Human Life*, ed. Kimberley Thachuk (Westport, Conn.: Praeger Security International, 2007), 166.
15. I use the word "cartel" in its vernacular connotation as a euphemism for criminal groups. In other Caribbean states, alternative names are used to convey the same concept. In Trinidad and Tobago and the Dominican Republic, the word "mafia" is common, while in Jamaica the word "posse" is used for the same purpose.
16. James Zackrison, "Smuggling and the Caribbean: Tainting Paradise throughout History," in *Transnational Threats: Smuggling and Trafficking in Arms, Drugs, and Human Life*, ed. Kimberley Thachuk (Westport, Conn.: Praeger Security International, 2007), 185.

17. U.S. Department of Homeland Security, *United States of America—Mexico: Bi-National Criminal Proceeds Study* (Washington, D.C.: U.S. Department of Homeland Security, 2010), 2-1.
18. Zackrison, "Smuggling and the Caribbean," 185.
19. Moisés Naím, *Illicit: How Smugglers, Traffickers, and Copycats Are Hijacking the Global Economy* (London: Arrow, 2007), 75.
20. Naím, 75.
21. Patrick Radden Keefe, "Cocaine Incorporated," *New York Times*, June 15, 2012, https://www.nytimes.com/2012/06/17/magazine/how-a-mexican-drug-cartel-makes-its-billions.html.
22. Keefe.
23. Keefe.
24. Gen. John Kelly, "Posture Statement of General John F. Kelly, United States Marine Corps Commander, United States Southern Command," testimony before the 113th Congress, House Armed Services Committee (Washington, D.C., February 26, 2014), 5.
25. Keefe, "Cocaine Incorporated."
26. Rear Adm. Charles Michel, "Written Statement of Rear Admiral Charles Michel Director Joint Interagency Task Force South (JIATF-South)," testimony before the Subcommittee on Border and Maritime Security, House Committee on Homeland Security Hearing on Border Security Threats to the Homeland: DHS's Response to Innovative Tactics and Techniques (Washington, D.C., June 19, 2012), 1.
27. Naím, *Illicit*, 27.
28. Keefe, "Cocaine Incorporated."
29. Danica Coto and David McFadden, "A Rising Tide of Drug Trafficking in the Caribbean," *San Diego Union-Tribune*, November 4, 2013, http://www.sandiegouniontribune.com/sdut-a-rising-tide-of-drug-trafficking-in-caribbean-2013nov04-story.html.
30. State Department, International Narcotics Control Strategy Report, vol. 1, *Drug and Chemical Control* (Washington, D.C.: State Department, March 2014), 8.
31. Kelly, "Posture Statement," (2014), 6.
32. Coto and McFadden, "A Rising Tide."
33. State Department, *International Narcotics Control Strategy Report*, 5.
34. State Department, 5.
35. State Department, 5.
36. Zackrison, "Smuggling and the Caribbean," 184–85.
37. Zackrison, 188–89.
38. Zackrison, 178.
39. Zackrison, 185.

40. Zackrison, 186.
41. Zackrison, 181, 185.
42. Millett, "Weak States and Porous Borders," 168.
43. Byron Ramirez, "Narco-Submarines: Drug Cartels' Innovative Technology," Center for International Maritime Security (CIMSEC) *NextWar* (blog), August 2, 2014, http://cimsec.org/narco-submarines-drug-cartels-innovative-technology/12314.
44. Ramirez; and André Hollis, "Narcoterrorism: A Definitional and Operational Transnational," in *Transnational Threats: Smuggling and Trafficking in Arms, Drugs, and Human Life*, ed. Kimberley Thachuk (Westport, Conn.: Praeger Security International, 2007), 27.
45. Keefe, "Cocaine Incorporated."
46. Hollis, "Narcoterrorism," 27.
47. Keefe, "Cocaine Incorporated."
48. Gen. John Kelly, "Posture Statement of General John F. Kelly, United States Marine Corps Commander, United States Southern Command," testimony before the 113th Congress, House Armed Services Committee (Washington, D.C., March 20, 2013), 14.
49. Michel, "Written Statement of Rear Admiral Charles Michel," 3.
50. Lizette Alvarez, "In Puerto Rico, Cocaine Gains Access to U.S.," *New York Times*, May 29, 2014, http://www.nytimes.com/2014/05/30/us/in-puerto-rico-cocaine-gains-access-to-us.html.
51. Kelly, "Posture Statement," (2014), 6.
52. Anthony Maingot, "The Challenge of the Corruption-Violence Connection," in *Caribbean Security in the Age of Terror: Challenge and Change*, ed. Ivelaw Griffith (Kingston, Jamaica: Ian Randle, 2004), 145.
53. Maingot, 145–46.
54. Maingot, 145–46.
55. Alvarez, "In Puerto Rico, Cocaine Gains Access to U.S."
56. Coto and McFadden, "A Rising Tide."
57. Mary Alice Young, "Dirty Money in Jamaica," *Journal of Money Laundering Control* 17, no. 3 (2014): 356.
58. Zackrison, "Smuggling and the Caribbean," 186.
59. Coto and McFadden, "A Rising Tide."
60. Coto and McFadden.
61. Arthur Hall, "Drug Bust at Sea—Almost $1b Worth of Cocaine Seized by JDF/JCF Team," *Gleaner* (Jamaica), April 17, 2016, http://jamaica-gleaner.com/article/lead-stories/20160417/drug-bust-sea-almost-1b-worth-cocaine-seized-jdfjcf-team.
62. State Department, *International Narcotics Control Strategy Report*, 5.
63. Griffith, "Probing Security Challenge," 35.
64. Young, "Dirty Money in Jamaica," 360.

65. Michel, "Written Statement of Rear Admiral Charles Michel," 2.
66. Ramirez, "Narco-Submarines."
67. Ramirez.
68. Maingot, "The Challenge of the Corruption-Violence Connection," 144.
69. State Department, *International Narcotics Control Strategy Report*, 48.
70. Michel, "Written Statement of Rear Admiral Charles Michel," 2.
71. Quoted in Meghann Myers, "In War with Drug Traffickers, Coast Guard Stretched Thin," *Navy Times*, October 18, 2014, https://www.navytimes.com/news/your-navy/2014/10/18/in-war-with-drug-traffickers-coast-guard-stretched-thin/.
72. Adm. Kurt Tidd, "Posture Statement of Admiral W. Kurt Tidd, Commander, United States Southern Command," testimony before the 115th Congress, Senate Armed Services Committee (Washington, D.C., April 6, 2017), 14.
73. Kelly, "Posture Statement," (2014), 43.
74. Quoted in Ramirez, "Narco-Submarines."
75. Kelly, "Posture Statement," (2014), 2.
76. Kelly, 3.
77. Tidd, "Posture Statement," 30.
78. Alvarez, "In Puerto Rico, Cocaine Gains Access to U.S."
79. Kelly, "Posture Statement," (2014), 17.
80. Michel, "Written Statement of Rear Admiral Charles Michel," 6.
81. State Department, *International Narcotics Control Strategy Report*, 44.
82. Kelly, "Posture Statement," (2014), 36.
83. Kelly, 43.
84. Kelly, 18.
85. Kelly, 43.
86. Kelly, 43.
87. "RFA Wave Knight Home in the Caribbean," Royal Navy website, last accessed January 25, 2016, http://www.royalnavy.mod.uk/news-and-latest-activity/news/2014/april/25/140425-rfa-wave-knight-home.
88. David McFadden, "Interpol-Led Operation Seizes Drugs in Caribbean," Associated Press, July 2, 2013, https://www.yahoo.com/news/interpol-led-operation-seizes-drugs-caribbean-200835751.html.
89. Zackrison, "Smuggling and the Caribbean," 184.
90. State Department, *International Narcotics Control Strategy Report*, 43.
91. Naím, *Illicit*, 274.
92. Quoted in Myers, "In War with Drug Traffickers."
93. Naím, *Illicit*, 14; see also Kelly, "Posture Statement," (2014), 6.
94. Zackrison, "Smuggling and the Caribbean," 180.
95. Naím, *Illicit*.
96. Naím, 77.

97. Naím, 14.
98. Maingot, "The Challenge of the Corruption-Violence Connection," 146.
99. "Drug Interdiction," United States Coast Guard website, last accessed October 21, 2018, https://www.gocoastguard.com/about-the-coast-guard/discover-our-roles-missions/drug-interdiction.

Chapter Four. Trafficking

1. Colvin Bishop and Oral Khan, "The Anti-Terrorism Capacity of Caribbean Security Forces," in *Caribbean Security in the Age of Terror: Challenge and Change*, ed. Ivelaw Griffith (Kingston, Jamaica: Ian Randle, 2004), 402.
2. Moisés Naím, *Illicit: How Smugglers, Traffickers, and Copycats Are Hijacking the Global Economy* (London: Arrow, 2007), 75.
3. Naím, 76.
4. Quoted in Naím, 66.
5. Zach Dyer, "Drug Traffickers Lure Costa Rica's Struggling Coastal Fishermen with Offers of Easy Money," *Tico Times* (Costa Rica), August 5, 2014, http://www.ticotimes.net/2014/08/05/drug-traffickers-lure-costa-ricas-struggling-coastal-fishermen-with-offers-of-easy-money.
6. Dyer.
7. Patrick Radden Keefe, "Cocaine Incorporated," *New York Times*, June 15, 2012, https://www.nytimes.com/2012/06/17/magazine/how-a-mexican-drug-cartel-makes-its-billions.html.
8. Ramesh Deosaran, "A Portrait of Crime in the Caribbean: Realities and Challenges," in *Caribbean Security in the Age of Terror: Challenge and Change*, ed. Ivelaw Griffith (Kingston, Jamaica: Ian Randle, 2004), 114.
9. Richard Millett, "Weak States and Porous Borders: Smuggling Along the Andean Ridge," in *Transnational Threats: Smuggling and Trafficking in Arms, Drugs, and Human Life*, ed. Kimberley Thachuk (Westport, Conn.: Praeger Security International, 2007), 171.
10. Millett, 169.
11. David McFadden, "With Murder Common, Jamaica Morgue Plans Stall," *Daily Mail*, June 21, 2014, https://www.dailymail.co.uk/wires/ap/article-2664286/With-murder-common-Jamaica-morgue-plans-stall.html.
12. Quoted in Mary Alice Young, "Dirty Money in Jamaica," *Journal of Money Laundering Control* 17, no. 3 (2014): 356.
13. McFadden, "With Murder Common."
14. U.S. Coast Guard (USCG), *Western Hemisphere Strategy* (Washington, D.C.: USCG, September 2014), 21.
15. Arabeska Sánchez, Leyla Diáz, and Matthias Nowak, "Firearms and Violence in Honduras," Small Arms Survey Research Notes, no. 39 (March 2014): 3.

16. Sánchez et al., 2.
17. Sánchez et al., 1.
18. Sánchez et al., 1.
19. Ivelaw Griffith, "Narcotics Arms Trafficking, Corruption and Governance in the Caribbean," *Journal of Money Laundering Control* 1, no. 2 (1997): 138.
20. Ivelaw Griffith, "Probing Security Challenge and Change in the Caribbean," in *Caribbean Security in the Age of Terror: Challenge and Change*, ed. Ivelaw Griffith (Kingston, Jamaica: Ian Randle, 2004), 33.
21. Gen. John Kelly, "Posture Statement of General John F. Kelly, United States Marine Corps Commander, United States Southern Command," testimony before the 113th Congress, House Armed Services Committee (Washington, D.C., February 26, 2014), 4.
22. State Department, *International Narcotics Control Strategy Report (INCSR)*, vol. 1, Drug and Chemical Control (Washington, D.C.: State Department, March 2014), 9.
23. Anthony Maingot, "The Challenge of the Corruption-Violence Connection," in *Caribbean Security in the Age of Terror: Challenge and Change*, ed. Ivelaw Griffith (Kingston, Jamaica: Ian Randle, 2004), 129.
24. David Kilcullen, *Out of the Mountains: The Coming Age of the Urban Guerrilla* (London: Hurst, 2013), 89–99.
25. State Department, *INCSR*, 191.
26. State Department, 139, 261.
27. Deosaran, "A Portrait of Crime in the Caribbean," 121.
28. "Recruit Training," Regional Police Training Centre (Barbados) website, last accessed January 26, 2016, https://web.archive.org/web/20160322181749/http://www.barbados-regional-police-training-center.gov.bb/page.aspx?page_id=37.
29. State Department, *INCSR*, 98.
30. State Department, 105, 139, 148, 166, 179.
31. Clifford Griffin, "Regional Law Enforcement Strategies in the Caribbean," in *Caribbean Security in the Age of Terror: Challenge and Change*, ed. Ivelaw Griffith (Kingston, Jamaica: Ian Randle, 2004), 488.
32. Deosaran, "A Portrait of Crime in the Caribbean," 113.
33. Griffith, "Probing Security Challenge and Change," 38.
34. Gen. John Kelly, "Posture Statement of General John F. Kelly, United States Marine Corps Commander, United States Southern Command," testimony before the 113th Congress, House Armed Services Committee (Washington, D.C., March 20, 2013), 8.
35. State Department, *Trafficking in Persons Report* (Washington, D.C.: State Department, 2014), https://www.state.gov/j/tip/rls/tiprpt/2014/index.htm, 29.

36. UN Office on Drugs and Crime (UNODC), *Global Report on Trafficking in Persons* (New York: UN Office on Drugs and Crime, December 2016), 90, https://www.unodc.org/documents/data-and-analysis/glotip/2016_Global_Report_on_Trafficking_in_Persons.pdf.
37. UNODC, *Global Report on Trafficking in Persons* (New York: UN Office on Drugs and Crime, 2012), 67; UNODC, *Global Report on Trafficking in Persons* (2016), 89.
38. Clare Ribando Seelke, *Trafficking in Persons in Latin America and the Caribbean*, CRS Report RL33200 (Washington, D.C.: Office of Congressional Information and Publishing, July 15, 2013), 3–4.
39. International Labour Organization (ILO), *Profits and Poverty: The Economics of Forced Labour* (Geneva: ILO, 2014), 13.
40. ILO, 17.
41. Lara Talsma, *Human Trafficking in Mexico and Neighboring Countries: A Review of Protection Approaches* (Geneva: UN High Commissioner for Refugees, June 2012), 5.
42. Seelke, *Trafficking in Persons*, 7; and Naím, *Illicit*, 88.
43. UNODC, *Global Report on Trafficking in Persons* (2016), 95.
44. UNODC, *Global Report on Trafficking in Persons* (2012), 14, 67.
45. UNODC, *Global Report on Trafficking in Persons* (2016), 96.
46. UNODC, 96.
47. UNODC, *Global Report on Trafficking in Persons* (2012), 64.
48. International Organization for Migration (IOM), *Exploratory Assessment of Trafficking in Persons in the Caribbean Region*, 2nd ed. (Washington, D.C.: IOM, 2010), 47.
49. Seelke, *Trafficking in Persons*, 6.
50. Seelke, 5.
51. "Child Labor in Latin America and the Caribbean," *International Labour Organization* website, accessed October 26, 2017, http://www.ilo.org/ipec/Regionsandcountries/latin-america-and-caribbean/lang—en/index.htm. Earlier estimates by the ILO place these figures nearly triple the rates currently being report. See Seelke, *Trafficking in Persons*, 5n21.
52. Seelke, *Trafficking in Persons*, 4.
53. Seelke, 1n3.
54. Millett, "Weak States and Porous Borders," 172.
55. Seelke, *Trafficking in Persons*, 5.
56. Millett, "Weak States and Porous Borders," 172.
57. Seelke, *Trafficking in Persons*, 5.
58. IOM, *Exploratory Assessment of Trafficking in Persons*, 24.
59. Naím, *Illicit*, 89.
60. Talsma, *Human Trafficking in Mexico*, 14.
61. Seelke, *Trafficking in Persons*, 5n24.
62. IOM, *Exploratory Assessment of Trafficking in Persons*, 41.

63. Francis Miko, "International Human Trafficking," in *Transnational Threats: Smuggling and Trafficking in Arms, Drugs, and Human Life*, ed. Kimberley Thachuk (Westport, Conn.: Praeger Security International, 2007), 43.
64. Seelke, *Trafficking in Persons*, 6, 7.
65. Naím, *Illicit*, 75; see also 76.
66. IOM, *Exploratory Assessment of Trafficking in Persons*, 18.
67. Naím, *Illicit*, 97–98.
68. Louise Shelley, "The Rise and Diversification of Human Smuggling and Trafficking in the United States," in *Transnational Threats: Smuggling and Trafficking in Arms, Drugs, and Human Life*, ed. Kimberley Thachuk (Westport, Conn.: Praeger Security International, 2007), 195.
69. Shelley, 195.
70. Millett, "Weak States and Porous Borders," 172.
71. Kelly, "Posture Statement," (2014), 4.
72. Kirk Semple, "Fleeing Gangs, Central American Families Surge toward U.S.," *New York Times*, November 12, 2016, https://www.nytimes.com/2016/11/13/world/americas/fleeing-gangs-central-american-families-surge-toward-us.html?_r=0.
73. Talsma, *Human Trafficking in Mexico*, 9.
74. Kelly, "Posture Statement," (2014), 8.
75. Adm. Kurt Tidd, "Posture Statement of Admiral W. Kurt Tidd, Commander, United States Southern Command Before the 115th Congress," testimony before the 115th Congress, Senate Armed Services Committee (Washington, D.C., April 6, 2017), 7.
76. Talsma, *Human Trafficking in Mexico*, 11.
77. Kelly, "Posture Statement," (2014), 7.
78. "US Coast Guard Alien Migrant Interdiction Operation Statistics," U.S. Coast Guard (website), last accessed October 22, 2018, https://migrantsatsea.org/us-coast-guard-alien-migrant-interdiction-operation-statistics/; and Associated Press, "Coast Guard Releases Migrant Data from Fiscal Year," *Washington Times*, October 5, 2014, http://www.washingtontimes.com/news/2014/oct/5/coast-guard-releases-migrant-data-from-fiscal-year/.
79. "US Coast Guard Alien Migrant Interdiction Operation Statistics"; and Associated Press, "Coast Guard Releases Migrant Data."
80. IOM, *Exploratory Assessment of Trafficking in Persons*, 25.
81. IOM, 63.
82. IOM, 54, 30.
83. IOM, 55.
84. Jared McCallister, "U.N. Agency Signals Alarm over Refugees Deaths after Another Maritime Disaster Claims Lives of Haitians," *Daily News*, December 1, 2013, http://www.nydailynews.com/new-york/sea-deaths-article-1.1533811.

Notes to Pages 105–110

85. Kelly, "Posture Statement," (2014), 8.
86. Bishop and Khan, "The Anti-Terrorism Capacity of Caribbean Forces," 398–99.
87. Jim Garamone, "Kelly: Southcom Keeps Watch on Ebola Situation," DoD News, October 8, 2014, https://dod.defense.gov/News/Article/Article/603408/kelly-southcom-keeps-watch-on-ebola-situation/.
88. Griffith, "Probing Security Challenge," 42.
89. Garamone, "Kelly."
90. USCG, *Western Hemisphere Strategy*, 22.
91. Millett, "Weak States and Porous Borders," 173.
92. IOM, *Exploratory Assessment of Trafficking in Persons*, 48, 68.
93. State Department, *INCSR*, 9.
94. Kelly, "Posture Statement," (2013), 10.
95. Kelly, "Posture Statement," (2014), 7.
96. State Department, *INCSR*, 13.
97. Griffith, "Probing Security Challenge," 36.
98. IOM, *Exploratory Assessment of Trafficking in Persons*, 27.
99. USCG, *Western Hemisphere Strategy*, 19.
100. IOM, *Exploratory Assessment of Trafficking in Persons*, 28.
101. For the 2016 figures, see an update to Seelke's report: Clare Riobando Seelke, *Trafficking in Persons in Latin America and the Caribbean*, CRS Report RL33200 (Washington, D.C.: Office of Congressional Information and Publishing, October 13, 2016), 11.
102. Kelly, "Posture Statement," (2014), 8.
103. Kelly, 20.
104. Kelly, 26.
105. Kelly, 27.
106. For more, see Joshua Tallis, "Other Than War: HA/DR and Geopolitics," Center for International Maritime Security (CIMSEC) *NextWar* (blog), March 28, 2016, http://cimsec.org/war-hadr-geopolitics/23591.
107. Rear Adm. Charles Michel, "Written Statement of Rear Admiral Charles Michel Director Joint Interagency Task Force South (JIATF-South)," testimony before the Subcommittee on Border and Maritime Security, House Committee on Homeland Security Hearing on Border Security Threats to the Homeland: DHS's Response to Innovative Tactics and Techniques (Washington, D.C., June 19, 2012), 1; and Tidd, "Posture Statement," 2.
108. Bishop and Khan, "The Anti-Terrorism Capacity of Caribbean Forces," 398.
109. Griffith, "Probing Security Challenge," 40.
110. André Hollis, "Narcoterrorism: A Definitional and Operational Transnational," in *Transnational Threats: Smuggling and Trafficking in Arms, Drugs,*

and Human Life, ed. Kimberley Thachuk (Westport, Conn.: Praeger Security International, 2007), 24–25.
111. Thomas Eyre, Affidavit in Support of Criminal Complaint, *United States of America v. Anthony Joseph Tracy*, Criminal Number 1:10-MJ-97, Eastern District of Virginia, Alexandria Division, http://www.investigativeproject.org/documents/case_docs/1300.pdf.
112. David McFadden, "Struggling Caribbean Islands Selling Citizenship," Associated Press, February 12, 2013, https://www.yahoo.com/news/struggling-caribbean-islands-selling-citizenship-170008926.html.
113. McFadden.
114. Arthur Brice, "Iran, Hezbollah Mine Latin America for Revenue, Recruits, Analysts Say," *CNN*, June 3, 2013, http://www.cnn.com/2013/06/03/world/americas/iran-latin-america/.
115. John Cope and Janie Hulse, "Hemispheric Response to Terrorism: A Call for Action," in *Caribbean Security in the Age of Terror: Challenge and Change*, ed. Ivelaw Griffith (Kingston, Jamaica: Ian Randle, 2004), 420.
116. Cope and Hulse, 417.
117. Brice, "Iran, Hezbollah Mine Latin America."
118. Kimberley Thachuk, "An Introduction to Transnational Threats," in *Transnational Threats: Smuggling and Trafficking in Arms, Drugs, and Human Life*, ed. Kimberley Thachuk (Westport, Conn.: Praeger Security International, 2007), 16.
119. Brice, "Iran, Hezbollah Mine Latin America."
120. Kelly, "Posture Statement," (2013), 13.
121. Cope and Hulse, "Hemispheric Response to Terrorism," 417.
122. "Profiles: Colombia's Armed Groups," *BBC*, August 29, 2013, http://www.bbc.com/news/world-latin-america-11400950.
123. Frank Cilluffo, "The Threat Posed from the Convergence of Organized Crime, Drug Trafficking, and Terrorism," testimony before the U.S. House Committee on the Judiciary Subcommittee on Crime, Washington, D.C., December 13, 2000, 4.
124. Naím, *Illicit*, 29.
125. Cilluffo, "Convergence of Organized Crime," 4; and Rhea Siers, "The Implications for U.S. National Security," in *Transnational Threats: Smuggling and Trafficking in Arms, Drugs, and Human Life*, ed. Kimberley Thachuk (Westport, Conn.: Praeger Security International, 2007), 214.
126. "Profiles: Colombia's Armed Groups."
127. Naím, *Illicit*, 70.
128. Thachuk, "An Introduction to Transnational Threats," 17.
129. Naím, *Illicit*, 79.
130. Siers, "The Implications for U.S. National Security," 218.
131. "Profiles: Colombia's Armed Groups."

132. Brice, "Iran, Hezbollah Mine Latin America."
133. Naím, *Illicit*, 137.
134. David A. Denny, "Terrorism, Drug Trafficking Inextricably Linked, U.S. Experts Say," *Washington File*, December 5, 2001, https://wfile.ait.org.tw/wf-archive/2001/011205/epf309.htm.
135. Denny.
136. Denny.

Chapter Five. Context and Conclusions
1. Moisés Naím, *Illicit: How Smugglers, Traffickers, and Copycats Are Hijacking the Global Economy* (London: Arrow, 2007), 141.
2. W. Andy Knight, "The Caribbean on the World Scene: Security Regimes, Instruments, and Actions," in *Caribbean Security in the Age of Terror: Challenge and Change*, ed. Ivelaw Griffith (Kingston, Jamaica: Ian Randle, 2004), 443.
3. Naím, *Illicit*, 141.
4. James Zackrison, "Smuggling and the Caribbean: Tainting Paradise throughout History," in *Transnational Threats: Smuggling and Trafficking in Arms, Drugs, and Human Life*, ed. Kimberley Thachuk (Westport, Conn.: Praeger Security International, 2007), 186.
5. Naím, *Illicit*, 147.
6. Zackrison, "Smuggling and the Caribbean," 185.
7. Patrick Radden Keefe, "Cocaine Incorporated," *New York Times*, June 15, 2012, www.nytimes.com/2012/06/17/magazine/how-a-mexican-drug-cartel-makes-its-billions.html.
8. U.S. Department of Homeland Security, *United States of America–Mexico: Bi-National Criminal Proceeds Study* (Washington, D.C.: U.S. Department of Homeland Security, 2010), 4-3.
9. Naím, *Illicit*, 78.
10. Naím, 149.
11. Mary Alice Young, "Dirty Money in Jamaica," *Journal of Money Laundering Control* 17, no. 3 (2014): 360.
12. Keefe, "Cocaine Incorporated."
13. Quoted in David A. Denny, "Terrorism, Drug Trafficking Inextricably Linked, U.S. Experts Say," *Washington File*, December 5, 2001, https://wfile.ait.org.tw/wf-archive/2001/011205/epf309.htm.
14. Keefe, "Cocaine Incorporated."
15. U.S. Department of Homeland Security, *United States of America–Mexico*, 4-1.
16. State Department, *International Narcotics Control Strategy Report (INCSR)*, vol. 1, Drug and Chemical Control (Washington, D.C.: State Department, March 2014), 3.

17. Shazeeda Ali, "A Tale of Two Regions: The Latin American and Caribbean Money-Laundering Connection," *Journal of Money Laundering Control* 2, no. 4 (1999): 312.
18. Young, "Dirty Money in Jamaica," 362.
19. John Cope and Janie Hulse, "Hemispheric Response to Terrorism: A Call for Action," in *Caribbean Security in the Age of Terror: Challenge and Change*, ed. Ivelaw Griffith (Kingston, Jamaica: Ian Randle, 2004), 419.
20. David Kilcullen, *Out of the Mountains: The Coming Age of the Urban Guerrilla* (London: Hurst, 2013), 89–99.
21. U.S. State Department Cable, American Embassy Kingston to Secretary of State, et al. 00682, OR 242332, May 2010, http://www.mattathiasschwartz.com/wp-content/uploads/2012/06/tny-cable.pdf.
22. Kilcullen, *Out of the Mountains*, 91.
23. Kimberley Thachuk, "An Introduction to Transnational Threats," in *Transnational Threats: Smuggling and Trafficking in Arms, Drugs, and Human Life*, ed. Kimberley Thachuk (Westport, Conn.: Praeger Security International, 2007), 10.
24. Ivelaw Griffith, "Narcotics Arms Trafficking, Corruption and Governance in the Caribbean," *Journal of Money Laundering Control* 1, no. 2 (1997), 142.
25. Griffith, 143.
26. Clifford Griffin, "Regional Law Enforcement Strategies in the Caribbean," in *Caribbean Security in the Age of Terror: Challenge and Change*, ed. Ivelaw Griffith (Kingston, Jamaica: Ian Randle, 2004), 500.
27. David McFadden, "St. Lucia PM: US Suspends Assistance to Police," Associated Press, August 21, 2013, https://finance.yahoo.com/news/st-lucia-pm-us-suspends-111754056.html.
28. Danica Coto, "Suriname Leader's Son Arrested on US Drug Charge," Associated Press, August 30, 2013, www.yahoo.com/news/suriname-leaders-son-arrested-us-drug-charge-190328268.html.
29. David McFadden, "Drug Charge for Virgin Islands Environment Officer," *Trentonian*, May 19, 2013, www.trentonian.com/drug-charge-for-virgin-islands-environment-officer/article_44c629ca-906f-55cc-817a-8fb69e136a10.html.
30. Quoted in Anthony Maingot, "The Challenge of the Corruption-Violence Connection," in *Caribbean Security in the Age of Terror: Challenge and Change*, ed. Ivelaw Griffith (Kingston, Jamaica: Ian Randle, 2004), 146.
31. Keefe, "Cocaine Incorporated."
32. Griffith, "Narcotics Arms Trafficking," 145.
33. Griffith, 145.
34. U.S. Coast Guard (USCG), *Western Hemisphere Strategy* (Washington, D.C.: U.S. Coast Guard, September 2014).

35. All quotes here are from USCG, 9.
36. USCG, 10.
37. USCG, 32.
38. USCG, 35.
39. Gen. John Kelly, "Posture Statement of General John F. Kelly, United States Marine Corps Commander, United States Southern Command," testimony before the 113th Congress, House Armed Services Committee (Washington, D.C., February 26, 2014), 29.
40. Meghann Myers, "In War with Drug Traffickers, Coast Guard Stretched Thin," *Navy Times*, October 18, 2014, https://www.navytimes.com/news/your-navy/2014/10/18/in-war-with-drug-traffickers-coast-guard-stretched-thin/.
41. Colvin Bishop and Oral Khan, "The Anti-Terrorism Capacity of Caribbean Forces," in *Caribbean Security in the Age of Terror: Challenge and Change*, ed. Ivelaw Griffith (Kingston, Jamaica: Ian Randle, 2004), 407.
42. Ivelaw Griffith, "Conclusion: Contending with Challenge, Coping with Change," in *Caribbean Security in the Age of Terror: Challenge and Change*, ed. Ivelaw Griffith (Kingston, Jamaica: Ian Randle, 2004), 516.
43. USCG, *Western Hemisphere Strategy*, 47.
44. Maingot, "The Challenge of the Corruption-Violence Connection," 137.
45. Zackrison, "Smuggling and the Caribbean," 180–81.
46. Richard Millett, "Weak States and Porous Borders: Smuggling along the Andean Ridge," in *Transnational Threats: Smuggling and Trafficking in Arms, Drugs, and Human Life*, ed. Kimberley Thachuk (Westport, Conn.: Praeger Security International, 2007), 166.
47. Zackrison, "Smuggling and the Caribbean," 181.
48. Trevor Munroe, "The Menace of Drugs," in *Caribbean Security in the Age of Terror: Challenge and Change*, ed. Ivelaw Griffith (Kingston, Jamaica: Ian Randle, 2004), 171. See also author interview with Capt. Mark Morris, USN (Ret.), former chief of the U.S. naval mission in Bogota, Colombia, January 22, 2015.
49. Kelly, "Posture Statement," 20.
50. Frank Cilluffo, "The Threat Posed from the Convergence of Organized Crime, Drug Trafficking, and Terrorism," testimony before the U.S. House Committee on the Judiciary Subcommittee on Crime, Washington, D.C., December 13, 2000, 2.
51. Raymond Milefsky, "Territorial Disputes and Regional Security in the Caribbean Basin," in *Caribbean Security in the Age of Terror: Challenge and Change*, ed. Ivelaw Griffith (Kingston, Jamaica: Ian Randle, 2004), 78.
52. Cilluffo, "Convergence of Organized Crime," 2.
53. Gen. John Kelly, "Posture Statement of General John F. Kelly, United States Marine Corps Commander, United States Southern Command,"

testimony before the 113th Congress, House Armed Services Committee (Washington, D.C., March 20, 2013), 10.
54. USCG, *Western Hemisphere Strategy*, 31.
55. Naím, *Illicit*, 81.
56. Naím, 67.
57. Bishop and Khan, "The Anti-Terrorism Capacity of Caribbean Forces," 407.
58. Isabel Jaramillo Edwards, "Coping with 9/11: State and Civil Society Responses," in *Caribbean Security in the Age of Terror: Challenge and Change*, ed. Ivelaw Griffith (Kingston, Jamaica: Ian Randle, 2004), 384.
59. Kelly, "Posture Statement," 12.
60. Millett, "Weak States and Porous Borders," 167.
61. Zackrison, "Smuggling and the Caribbean," 181.
62. Millett, "Weak States and Porous Borders," 167.
63. Millett, 167.
64. Kelly, "Posture Statement (2013)," 2.
65. Ivelaw Griffith, "Probing Security Challenge and Change in the Caribbean," in *Caribbean Security in the Age of Terror: Challenge and Change*, ed. Ivelaw Griffith (Kingston, Jamaica: Ian Randle, 2004), 33.
66. Zackrison, "Smuggling and the Caribbean," 191.
67. State Department, *INCSR*, 179, 237.
68. State Department, 211.
69. Zackrison, "Smuggling and the Caribbean," 190. Emphasis added.
70. Zackrison, 190.
71. Juanita Darling, "Submarine Links Colombian Drug Traffickers with Russian Mafia," *Los Angeles Times*, November 10, 2000, http://articles.latimes.com/2000/nov/10/news/mn-49908.
72. Zackrison, "Smuggling and the Caribbean," 191.
73. Edwards, "Coping with 9/11," 377.

Chapter Six. Piracy and Perception in the Gulf of Guinea

1. Chatham House, *Maritime Security in the Gulf of Guinea* (London: Chatham House, March 2013), 1; and Freedom Onuoha, "The Geo-Strategy of Oil in the Gulf of Guinea: Implications for Regional Stability," *Journal of Asian and African Studies* 45, no. 3 (2010): 370.
2. Chatham House, *Maritime Security in the Gulf of Guinea*, 1.
3. Adeniyi Adejimi Osinowo, "Combating Piracy in the Gulf of Guinea," *Africa Security Brief*, no. 30 (February 2015): 1.
4. Ali Kamal-Deen, "The Anatomy of the Gulf of Guinea Piracy," *Naval War College Review* 68, no. 1 (Winter 2015): 94.
5. International Crisis Group (ICG), "The Gulf of Guinea: The New Danger Zone," *Crisis Group Africa Report*, no. 195 (December 2012): 2n7.

6. Onuoha, "The Geo-Strategy of Oil in the Gulf of Guinea," 370.
7. Kamal-Deen, "The Anatomy of the Gulf of Guinea Piracy," 94.
8. Lisa Otto, "Westward Ho! The Evolution of Maritime Piracy in Nigeria," *Portuguese Journal of Social Science* 13, no. 3 (2014): 318.
9. Francois Vreÿ, "Maritime Aspects of Illegal Oil-Bunkering in the Niger Delta," *Australian Journal of Maritime and Ocean Affairs* 4, no. 4 (2012): 110.
10. Vreÿ, 111.
11. David Kilcullen, *Out of the Mountains: The Coming Age of the Urban Guerrilla* (London: Hurst, 2013).
12. ICG, "The Gulf of Guinea," 4.
13. Raymond Gilpin, "Enhancing Maritime Security in the Gulf of Guinea," *Strategic Insights* 6, no. 1 (January 2007): 1, https://calhoun.nps.edu/handle/10945/11174. Vreÿ identifies 2008–10 as the central years of this reorientation. See Francois Vreÿ, "African Maritime Security: A Time for Good Order at Sea," *Australian Journal of Maritime and Ocean Affairs* 2, no. 4 (2010): 121.
14. Kamal-Deen, "The Anatomy of the Gulf of Guinea Piracy," 107.
15. Onuoha, "The Geo-Strategy of Oil in the Gulf of Guinea," 374.
16. Stephen Ellis, "West Africa's International Drug Trade," *African Affairs* 108, no. 431 (2009): 185, http://dx.doi.org/10.1093/afraf/adp017.
17. Quoted in "The Navy of Ghana—Making the Gulf of Guinea's Coastal Waters the Safest in the Region," *Military Technology*, October 2014, 58.
18. Gilpin, "Enhancing Maritime Security in the Gulf of Guinea."
19. Ellis, "West Africa's International Drug Trade," 172; and UN Office on Drugs and Crime (UNODC), *Cocaine Trafficking in West Africa: The Threat to Stability and Development (with Special Reference to Guinea-Bissau)* (New York: UNODC, December 2007), 1.
20. UNODC, *Cocaine Trafficking in West Africa*, 3.
21. Casey Wilander, "Illicit Confluence: The Intersection of Cocaine and Illicit Timber in the Amazon," *Small Wars Journal*, October 16, 2017, http://smallwarsjournal.com/jrnl/art/illicit-confluences-the-intersection-of-cocaine-and-illicit-timber-in-the-amazon#_ednref57.
22. Ellis, "West Africa's International Drug Trade," 191.
23. Moisés Naím, *Illicit: How Smugglers, Traffickers, and Copycats Are Hijacking the Global Economy* (London: Arrow, 2007), 194; and Ellis, "West Africa's International Drug Trade," 184, 190.
24. Ellis, "West Africa's International Drug Trade," 191.
25. Chatham House, *Maritime Security in the Gulf of Guinea*, 17–18.
26. Naím, *Illicit*, 29.
27. Andre Le Sage, "Africa's Irregular Security Threats: Challenges for U.S. Engagement," *Strategic Forum*, no. 255 (May 2010): 5; and UNODC, *Cocaine Trafficking in West Africa*, 20.

28. Vreÿ, "African Maritime Security," 125.
29. Le Sage, "Africa's Irregular Security Threats," 6.
30. Le Sage, 7; and Audra Grant, "Smuggling and Trafficking in Africa," in *Transnational Threats: Smuggling and Trafficking in Arms, Drugs, and Human Life*, ed. Kimberley Thachuk (Westport, Conn.: Praeger Security International, 2007), 117.
31. Le Sage, "Africa's Irregular Security Threats," 7.
32. Ellis, "West Africa's International Drug Trade," 187.
33. Naím, *Illicit*, 73.
34. Grant, "Smuggling and Trafficking in Africa," 117.
35. Ellis, "West Africa's International Drug Trade," 193.
36. Le Sage, "Africa's Irregular Security Threats," 7.
37. UN Office on Drugs and Crime, *The Drug Problem and Organized Crime, Illicit Financial Flows, Corruption and Terrorism* (New York: UNODC, May 2017), 11.
38. Vreÿ, "African Maritime Security," 125.
39. Ellis, "West Africa's International Drug Trade," 182–83.
40. Ellis, 192–93.
41. Grant, "Smuggling and Trafficking in Africa," 119.
42. Vreÿ, "African Maritime Security," 123.
43. UNODC, *Cocaine Trafficking in West Africa*, 1.
44. Le Sage, "Africa's Irregular Security Threats," 6.
45. Le Sage, 7.
46. Ellis, "West Africa's International Drug Trade," 192.
47. European Union, *Fact Sheet: EU Strategy on the Gulf of Guinea* (Brussels: European Union, March 17, 2014), 2.
48. Vreÿ, "Maritime Aspects of Illegal Oil-Bunkering," 111.
49. State Department, *International Narcotics Control Strategy Report*, vol. 1, Drug and Chemical Control (Washington, D.C.: U.S. Department of State, March 2014), 10.
50. Geoffrey Till, *Seapower: A Guide for the Twenty-First Century*, 2nd ed. (New York: Routledge, 2009), 306.
51. State Department, *INCSR*, 10.
52. UNODC, *Cocaine Trafficking in West Africa*, 8.
53. UNODC, 11.
54. Ellis, "West Africa's International Drug Trade," 194.
55. State Department, *INCSR*, 254.
56. Ellis, "West Africa's International Drug Trade," 181.
57. UNODC, *Cocaine Trafficking in West Africa*, 1.
58. Noted in Otto, "Westward Ho!," 317.
59. Quoted in Ellis, "West Africa's International Drug Trade," 194.
60. Le Sage, "Africa's Irregular Security Threats," 2.
61. Grant, "Smuggling and Trafficking in Africa," 120.

62. Naím, *Illicit*, 52.
63. Onuoha, "The Geo-Strategy of Oil," 397.
64. Naím, *Illicit*, 54.
65. Grant, "Smuggling and Trafficking in Africa," 121.
66. Lauren Ploch, *Africa Command: U.S. Strategic Interests and the Role of the U.S. Military in Africa*, CRS Report RL34003, U.S. Library of Congress, Congressional Research Service (Washington, D.C.: Office of Congressional Information and Publishing, October 2, 2009), 17; and Naím, *Illicit*, 29.
67. Grant, "Smuggling and Trafficking in Africa," 122.
68. See Naím, *Illicit*, 150.
69. Le Sage, "Africa's Irregular Security Threats," 2.
70. Chatham House, *Maritime Security in the Gulf of Guinea*, 3. In Ghana, Equatorial Guinea, and Sao Tome and Principe, for example, fish comprise roughly 60 percent of protein consumption. See Gilpin, "Enhancing Maritime Security," 2.
71. International Labour Organization, *Caught at Sea: Forced Labour and Trafficking in Fisheries* (Geneva: International Labour Organization, 2013), 5.
72. Osinowo, "Combating Piracy in the Gulf of Guinea," 1–2.
73. European Union, *Fact Sheet*, 2.
74. U.S. State Department Office to Monitor and Combat Trafficking in Persons, *The Intersection between Environmental Degradation and Human Trafficking* (Washington, D.C.: U.S. State Department, June 2014), 2.
75. Osinowo, "Combating Piracy in the Gulf of Guinea," 7.
76. Otto, "Westward Ho!," 318.
77. See, for example, Mariama Sow, "Figures of the Week: Piracy and Illegal Fishing in Somalia," *Africa in Focus* (Brookings), April 12, 2017, https://www.brookings.edu/blog/africa-in-focus/2017/04/12/figures-of-the-week-piracy-and-illegal-fishing-in-somalia/; and Ishaan Tharoor, "How Somalia's Fishermen Became Pirates," *Time*, April 18, 2009, http://content.time.com/time/world/article/0,8599,1892376,00.html.
78. Chatham House, *Maritime Security in the Gulf of Guinea*, 16.
79. Grant, "Smuggling and Trafficking in Africa," 125.
80. Grant, 126.
81. Murphy, "The Troubled Waters of Africa," 75.
82. Otto, "Westward Ho!," 319.
83. Quoted in Jon Gambrell and Associated Press, "Oil Bunkering Threatens Nigeria's Economy, Environment," *Washington Post*, July 20, 2013, https://www.washingtonpost.com/national/oil-bunkering-threatens-nigerias-economy-environment/2013/07/18/e38cb4a0-e273-11e2-aef3-339619eab080_story.html.
84. Vreÿ, "Maritime Aspects of Illegal Oil-Bunkering," 114.

85. Vreÿ, 111.
86. Murphy, "The Troubled Waters of Africa," 74.
87. Otto, "Westward Ho!," 318.
88. Kamal-Deen, "The Anatomy of the Gulf of Guinea Piracy," 97.
89. Otto, "Westward Ho!," 318.
90. Kamal-Deen, "The Anatomy of the Gulf of Guinea Piracy," 98.
91. Otto, "Westward Ho!," 318.
92. Murphy, "The Troubled Waters of Africa," 74.
93. Le Sage, "Africa's Irregular Security Threats," 2.
94. Threats to sovereignty do not only come from informal actors. There is a major industry for security contractors in Africa, as elsewhere. In Cameroon, vigilantes serve lower-income communities while nearly two hundred official private security firms serve the country's wealthy. A similar phenomenon was recorded in Nigeria, where a growth in private security paralleled the expansion of organizations such as the Vigilante Groups of Nigeria, which in one Nigerian city served more than four thousand households (for a fee). See Naím, *Illicit*, 61.
95. Quote in Rebecca Jones, *State Failure and Extra-Legal Justice: Vigilante Groups, Civil Militias and the Rule of Law in West Africa* (Geneva: UN High Commissioner for Refugees, October 2008), 1.
96. Le Sage, "Africa's Irregular Security Threats," 8. See, for example, Amnesty International, *"Welcome to Hell Fire": Torture and Other Ill-Treatment in Nigeria* (London: Amnesty International, 2014); and "Nigeria 'Uses Torture Officers to Extract Confessions,'" BBC, September 18, 2014, http://www.bbc.com/news/world-africa-29254500?utm_source=Sailthru&utm_medium=email&utm_term=%252AMorning%20Brief&utm_campaign=2014_MorningBrief%209.18.14.
97. Jones, *State Failure and Extra-Legal Justice*, 3.
98. Murphy, "The Troubled Waters of Africa," 77.
99. Chatham House, *Maritime Security in the Gulf of Guinea*, 18.
100. Grant, "Smuggling and Trafficking in Africa," 124.
101. Le Sage, "Africa's Irregular Security Threats," 6.
102. UNODC, *Global Report on Trafficking in Persons* (New York: UNODC, December 2012), http://www.unodc.org/documents/data-and-analysis/glotip/Trafficking_in_Persons_2012_web.pdf, 78.
103. UNODC, *Cocaine Trafficking in West Africa*, 28; UNODC, *Global Report on Trafficking in Persons*, 78.
104. International Labour Organization (ILO), *Profits and Poverty: The Economics of Forced Labour* (Geneva: International Labour Organization, 2014), 7, 14.
105. UNODC, *Global Report on Trafficking in Persons*, 76.
106. UNODC, 77.

107. UNODC, 74.
108. ILO, *Profits and Poverty*, 17.
109. UNODC, *Global Report on Trafficking in Persons*, 79.
110. ILO, *Profits and Poverty*, 13.
111. Norimitsu Onishi, "Clashes Erupt as Liberia Sets an Ebola Quarantine," *New York Times*, August 20, 2014, http://www.nytimes.com/2014/08/21/world/africa/ebola-outbreak-liberia-quarantine.html?emc=edit_th_20140821&nl=todaysheadlines&nlid=58153314&_r=0.
112. Le Sage, "Africa's Irregular Security Threats," 2.
113. International Organization for Migration (IOM), *World Migration Report 2013: Migrant Well-Being and Development* (Geneva: IOM, 2013), 59.
114. Eero Tepp, "The Gulf of Guinea: Military and Non-Military Ways of Combatting Piracy," *Baltic Security and Defence* 14, no. 1 (2012): 191.
115. Murphy, "The Troubled Waters of Africa," 83.
116. UNODC, *Cocaine Trafficking in West Africa*, 30.
117. Tepp, "The Gulf of Guinea," 191.
118. "Nigeria 'Uses Torture Officers to Extract Confessions'."
119. Murphy, "The Troubled Waters of Africa," 81.
120. Dirk Steffen, "Risks in Contracting Government Security Forces in the Gulf of Guinea," Center for International Maritime Security (CIMSEC) *NextWar* (blog), September 16, 2014, http://cimsec.org/troubled-waters-2-risks-contracting-government-security-forces-gulf-guinea/13016.
121. Tepp, "The Gulf of Guinea," 201. See also Chatham House, *Maritime Security in the Gulf of Guinea*, 23.
122. ICG, "The Gulf of Guinea," 8.
123. Steffen, "Risks in Contracting Government Security Forces."
124. Vreÿ, "Maritime Aspects of Illegal Oil-Bunkering," 113.
125. Tepp, "The Gulf of Guinea," 201.
126. Le Sage, "Africa's Irregular Security Threats," 7.
127. Le Sage, 7.
128. Le Sage, 7.
129. Kamal-Deen, "The Anatomy of the Gulf of Guinea Piracy," 107.
130. Otto, "Westward Ho!," 315.
131. UN Convention on the Law of the Sea, December 10, 1982, http://www.un.org/Depts/los/convention_agreements/convention_overview_convention.htm.
132. Otto, "Westward Ho!," 316.
133. Tepp, "The Gulf of Guinea," 188.
134. Tepp, 185.
135. Freedom Onuoha, "Piracy and Maritime Security off the Horn of Africa: Connections, Causes, and Concerns," *African Security* 3, no. 4 (2010): 193.
136. Otto, "Westward Ho!," 316.
137. Otto, 316.

138. Kamal-Deen, "The Anatomy of the Gulf of Guinea Piracy," 94.
139. Murphy, "The Troubled Waters of Africa," 77.
140. Francois Vreÿ, "Bad Order at Sea: From the Gulf of Aden to the Gulf of Guinea," *African Security Review* 18, no. 3 (September 2009): 23; and Kamal-Deen, "The Anatomy of the Gulf of Guinea Piracy," 95.
141. Dirk Steffen, "Challenging the Myths of Pirate Violence," *Marine Link*, September 2, 2014, http://www.marinelink.com/news/challenging-violence376188.aspx.
142. Vreÿ, "African Maritime Security," 123.
143. Rick Noack, "Why Nigeria's Election Year May See a Spike in Pirate Attacks," *Washington Post*, October 14, 2014, https://www.washingtonpost.com/news/worldviews/wp/2014/10/14/why-nigerias-election-year-may-see-a-spike-in-pirate-attacks/.
144. Tepp, "The Gulf of Guinea," 184.
145. Otto, "Westward Ho!," 316.
146. Kamal-Deen, "The Anatomy of the Gulf of Guinea Piracy," 95.
147. Based on analysis of numbers from the International Maritime Bureau's *Piracy and Armed Robbery against Ships* reports from 2005, 2010, 2015, and 2016 covering years from 1994 to 2016. http://www.allaboutshipping.co.uk/wp-content/uploads/2017/01/2016-Annual-IMB-Piracy-Report-ABRIDGED.pdf.
148. ICG, "The Gulf of Guinea," 9.
149. Kamal-Deen, "The Anatomy of the Gulf of Guinea Piracy," 101.
150. Le Sage, "Africa's Irregular Security Threats," 5.
151. ICG, "The Gulf of Guinea," 10.
152. IMB, "Piracy and Armed Robbery against Ships," 5.
153. Kamal-Deen, "The Anatomy of the Gulf of Guinea Piracy," 98.
154. Osinowo, "Combating Piracy in the Gulf of Guinea," 2.
155. Kamal-Deen, "The Anatomy of the Gulf of Guinea Piracy," 95.
156. Donald Puchala, "Of Pirates and Terrorists: What Experience and History Teach," *Contemporary Security Policy* 26, no. 1 (April 2005): 2.
157. Osinowo, "Combating Piracy in the Gulf of Guinea," 2.
158. Kamal-Deen, "The Anatomy of the Gulf of Guinea Piracy," 93.
159. ICG, "The Gulf of Guinea," 15.
160. Based on analysis of numbers from the International Maritime Bureau's *Piracy and Armed Robbery against Ships* reports from 2005, 2010, 2015, and 2016 covering years from 1994 to 2016.
161. Osinowo, "Combating Piracy in the Gulf of Guinea," 3; and Kamal-Deen, "The Anatomy of the Gulf of Guinea Piracy," 103.
162. Kamal-Deen, "The Anatomy of the Gulf of Guinea Piracy," 95; and numbers based on analysis of numbers from the International Maritime Bureau's *Piracy and Armed Robbery against Ships* reports from 2005, 2010, 2015, and 2016 covering years from 1994 to 2016.

163. Steffen, "Challenging the Myths of Pirate Violence."
164. ICG, "The Gulf of Guinea," 13.
165. Tepp, "The Gulf of Guinea," 187.
166. Kamal-Deen, "The Anatomy of the Gulf of Guinea Piracy," 100.
167. Noack, "Why Nigeria's Election Year."
168. ICG, "The Gulf of Guinea," 11.
169. Kamal-Deen, "The Anatomy of the Gulf of Guinea Piracy," 104.
170. Tepp, "The Gulf of Guinea," 187.
171. Osinowo, "Combating Piracy in the Gulf of Guinea"; and Otto, "Westward Ho!"
172. Kamal-Deen, "The Anatomy of the Gulf of Guinea Piracy," 102.
173. Murphy, "The Troubled Waters of Africa," 73.
174. Otto, "Westward Ho!," 320.
175. Tepp, "The Gulf of Guinea," 188.
176. Kamal-Deen, "The Anatomy of the Gulf of Guinea Piracy," 101. While this violence is widely cited, Steffen disputes its treatment as dogma and dives into such claims in greater detail. See Steffen, "Challenging the Myths of Pirate Violence."
177. Gilpin, "Enhancing Maritime Security in the Gulf of Guinea," 2–3.
178. Kamal-Deen, "The Anatomy of the Gulf of Guinea Piracy," 95, 97.
179. Tepp, "The Gulf of Guinea," 194; and Kamal-Deen, "The Anatomy of the Gulf of Guinea Piracy," 101.
180. Chatham House, *Maritime Security in the Gulf of Guinea*, 21.
181. Stephen Starr, "Maritime Piracy on the Rise in West Africa," *CTC Sentinel* 7, no. 4 (April 2014): 24.
182. Onuoha, "Piracy and Maritime Security," 204.
183. Kamal-Deen, "The Anatomy of the Gulf of Guinea Piracy," 106.
184. Quoted in Murphy, "The Troubled Waters of Africa," 73–74.
185. Ploch, *Africa Command*, 24.
186. Ploch, 25.
187. Ploch, 6.
188. Chatham House, *Maritime Security in the Gulf of Guinea*, 7.
189. Chatham House, 34.
190. Osinowo, "Combating Piracy in the Gulf of Guinea," 4.
191. Code of Conduct Concerning the Repression of Piracy, Armed Robbery against Ships, and Illicit Maritime Activity in West and Central Africa (Yaounde, Cameroon: June 25, 2013): 1.
192. ICG, "The Gulf of Guinea," 21; and Osinowo, "Combating Piracy in the Gulf of Guinea," 4.
193. ICG, "The Gulf of Guinea," 21.
194. Tepp, "The Gulf of Guinea," 197.
195. Tepp, 196.

196. European Commission, "New EU Initiative to Combat Piracy in the Gulf of Guinea," Press Release, January 10, 2013, http://europa.eu/rapid/press-release_IP-13-14_en.htm.
197. Kamal-Deen, "The Anatomy of the Gulf of Guinea Piracy," 109.
198. Yao Jianjing, "Chinese Soldiers to Join Anti-Piracy Efforts in the Gulf of Guinea," *Xinhua*, July 29, 2016.
199. ICG, "The Gulf of Guinea," 3.
200. Chatham House, *Maritime Security in the Gulf of Guinea*, 31.
201. Kamal-Deen, "The Anatomy of the Gulf of Guinea Piracy," 109.
202. Ploch, *Africa Command*, 21.
203. Ploch, 5–6.
204. Starr, "Maritime Piracy on the Rise in West Africa," 25.
205. Paul Pryce, "Gabon's Growing Navy," Center for International Maritime Security (CIMSEC) *NextWar* (blog), December 17, 2014, http://cimsec.org/gabons-growing-navy/13975.
206. ICG, "The Gulf of Guinea," 5.
207. Murphy, "The Troubled Waters of Africa," 69.
208. Chatham House, *Maritime Security in the Gulf of Guinea*, 20.
209. Otto, "Westward Ho!," 322; emphasis is added.
210. Tepp, "The Gulf of Guinea," 192.
211. Tepp, 192.
212. Murphy, "The Troubled Waters of Africa," 69.
213. Vreÿ, "Bad Order at Sea," 27.
214. Quoted in Noack, "Why Nigeria's Election Year."
215. Osinowo, "Combating Piracy in the Gulf of Guinea," 5.

Chapter Seven. Evolving Security Conceptions in the Straits

1. Nazery Khalid, "With a Little Help from My Friends: Maritime Capacity-Building Measures in the Straits of Malacca," *Contemporary Southeast Asia* 31, no. 3 (2009): 424–25.
2. Khalid, 425.
3. J. Ashley Roach, "Enhancing Maritime Security in the Straits of Malacca and Singapore," *Journal of International Affairs* 59, no. 1 (Fall/Winter 2005): 97–98.
4. Roach, 100.
5. Sheldon Simon, "Safety and Security in the Malacca Straits: The Limits of Collaboration," *Asian Security* 7, no. 1 (2011): 27.
6. Upward of 80 percent of the oil and gas imports of China, Japan, Taiwan, and South Korea pass through the Strait of Malacca. See Yann-huei Song, "Security in the Strait of Malacca and the Regional Maritime Security Initiative: Responses to the U.S. Proposal," *International Law Studies* 83 (2009): 98. Moreover, half of the world's oil overall passes through the

Straits of Malacca and Singapore. See Victor Huang, "Building Maritime Security in Southeast Asia: Outsiders Not Welcome?," *Naval War College Review* 61, no. 1 (Winter 2008): 87. Finally, two-thirds of the world's natural gas shipments also pass through the Malacca Strait annually as well as a quarter of all global trade. See Gal Luft and Anne Korin, "Terrorism Goes to Sea," *Foreign Policy* 83, no. 6 (November–December 2004): 67.

7. David Rosenberg, "The Maritime Borderlands: Terrorism, Piracy, Pollution, and Poaching in the South China Sea," in *The Borderlands of Southeast Asia: Geopolitics, Terrorism, and Globalization*, ed. James Clad, Sean McDonald, and Bruce Vaughn (Washington, D.C.: National Defense University Press, 2011), 110.
8. Rosenberg, 111.
9. John Bradford, "Southeast Asian Maritime Security in the Age of Terror: Threats, Opportunity, and Changing the Course Forward," Nanyang Technological University RSIS working paper, no. 75 (2005): 1.
10. Song, "Security in the Strait of Malacca," 98.
11. Sam Bateman, "The Future Maritime Security Environment in Asia: A Risk Assessment Approach," *Contemporary Southeast Asia* 37, no. 1 (2015): 59–60.
12. Justin Hastings, "Geographies of State Failure and Sophistication in Maritime Piracy Hijackings," *Political Geography* 28, no. 4 (2009): 220.
13. Ted Kemp, "Crime on the High Seas: The World's Most Pirated Waters," *CNBC*, September 15, 2014, http://www.cnbc.com/2014/09/15/worlds-most-pirated-waters.html.
14. Carolin Liss, "New Actors and the State: Addressing Maritime Security Threats in Southeast Asia," *Contemporary Southeast Asia* 35, no. 2 (2013): 148.
15. Quoted in Nani Afrida, "Narcotics Agency, Navy Stay Vigilant on Illegal Ports," *Jakarta Post* (Indonesia), March 13, 2015, http://www.thejakartapost.com/news/2015/03/13/narcotics-agency-navy-stay-vigilant-illegal-ports.html.
16. Bateman, "The Future Maritime Security Environment in Asia," 61–64.
17. Bateman, 57–61.
18. Aditi Chatterjee, "Non-Traditional Maritime Security Threats in the Indian Ocean Region," *Maritime Affairs* 10, no. 2 (Winter 2014): 86.
19. Alan Dupont, "Transnational Crime, Drugs, and Security in East Asia," *Asian Survey* 39, no. 3 (May–June 1999): 451.
20. Ralf Emmers, "The Threat of Transnational Crime in Southeast Asia: Drug Trafficking, Human Smuggling and Trafficking, and Sea Piracy," *UNISCI Discussion Papers*, May 2003, 4.
21. State Department, "Southeast Asia Maritime Law Enforcement Initiative," *Fact Sheet*, April 10, 2015, http://www.state.gov/r/pa/pl/240798.htm.

22. Bateman, "The Future Maritime Security Environment in Asia," 60.
23. Rosenberg, "The Maritime Borderlands," 116.
24. Prashanth Parameswaran, "Explaining Indonesia's 'Sink the Vessel' Policy under Jokowi," *Diplomat*, January 13, 2015, http://thediplomat.com/2015/01/explaining-indonesias-sink-the-vessels-policy-under-jokowi/.
25. Bateman, "The Future Maritime Security Environment in Asia," 63.
26. Roach, "Enhancing Maritime Security," 100.
27. Chatterjee, "Non-Traditional Maritime Security Threats," 85.
28. John Bradford, "Shifting the Tides against Piracy in Southeast Asian Waters," *Asian Survey* 48, no. 3 (May–June 2008), 490.
29. Bateman, "The Future Maritime Security Environment in Asia," 59.
30. Liss, "New Actors and the State," 149.
31. As in the case of Gulf of Guinea scholarship, much of the literature on Southeast Asian piracy recognizes the limitations of UN Convention on the Law of the Sea's definition of piracy (Tammy Sittnick, "State Responsibility and Maritime Terrorism in the Strait of Malacca: Persuading Indonesia and Malaysia to Take Additional Steps to Secure the Strait," *Pacific Rim Law and Policy Journal* 14, no. 3 [2005]: 756) and instead defines piracy and armed robbery collectively, in line with the definition of the International Maritime Bureau. See Adam Young and Mark Valencia, "Conflation of Piracy and Terrorism in Southeast Asia: Rectitude and Utility," *Contemporary Southeast Asia* 25, no. 2 (August 2003): 270; Ralf Emmers, "ASEAN and the Securitization of Transnational Crime in Southeast Asia," *Pacific Review* 16, no. 3 (2003): 7; and Neil Renwick and Jason Abbott, "Piratical Violence and Maritime Security in Southeast Asia," *Security Dialogue* 30, no. 2 (June 1999): 183.
32. Chatterjee, "Non-Traditional Maritime Security Threats," 80.
33. Simon, "Safety and Security in the Malacca Straits," 27.
34. Hastings, "Geographies of State Failure," 215.
35. Chatterjee, "Non-Traditional Maritime Security Threats," 81.
36. Young and Valencia, "Conflation of Piracy and Terrorism in Southeast Asia," 271.
37. Bradford, "Shifting the Tides," 473.
38. ICC International Maritime Bureau (ICC IMB), *Piracy and Armed Robbery against Ships Report for the Period 1 January–31 December 2014* (London: ICC IMB, January 2015), 2.
39. Kemp, "Crime on the High Seas."
40. Catherine Zara Raymond, "Piracy and Armed Robbery in the Malacca Strait," *Naval War College Review* 62, no. 3 (Summer 2009): 32.
41. Sittnick, "State Responsibility," 753.
42. Yun Yun Teo, "Target Malacca Straits: Maritime Terrorism in Southeast Asia," *Studies in Conflict and Terrorism* 30, no. 6 (2007): 541.

43. Raymond, "Piracy and Armed Robbery," 36; and Simon, "Safety and Security in the Malacca Straits," 35.
44. Raymond, "Piracy and Armed Robbery," 37–38.
45. Bradford, "Shifting the Tides," 484.
46. Simon, "Safety and Security in the Malacca Straits," 37.
47. Simon, 32.
48. Simon, 30.
49. Song, "Security in the Strait of Malacca," 119.
50. Teo, "Target Malacca Straits," 554.
51. Teo, 543; and Huang, "Building Maritime Security in Southeast Asia," 90.
52. Simon, "Safety and Security in the Malacca Straits," 32.
53. Teo, "Target Malacca Straits," 543.
54. Song, "Security in the Strait of Malacca," 124.
55. Teo, "Target Malacca Straits," 547.
56. Song, "Security in the Strait of Malacca," 99, 120; and "MISC Hands over Vessel to Royal Malaysian Navy," *Hellenic Shipping News Worldwide*, January 23, 2016, http://www.hellenicshippingnews.com/misc-hands-over-vessel-to-royal-malaysian-navy/.
57. Simon, "Safety and Security in the Malacca Straits," 32, 36.
58. Khalid, "With a Little Help from My Friends," 432.
59. Simon, "Safety and Security in the Malacca Straits," 36; and Khalid, "With a Little Help from My Friends," 432.
60. Teo, "Target Malacca Straits," 432.
61. John Bradford, "The Maritime Strategy of the United States: Implications for Indo-Pacific Sea Lanes," *Contemporary Southeast Asia* 33, no. 2 (August 2011): 196–97.
62. Bradford, 195–96.
63. State Department, "Southeast Asia Maritime Law Enforcement Initiative."
64. Roach, "Enhancing Maritime Security," 107.
65. Bradford, "Shifting the Tides," 482.
66. Bradford, 478; Raymond, "Piracy and Armed Robbery," 37; and Donna Nincic, "Trends in Modern Piracy: Cycles, Geographical Shifts, and Predicting the Next 'Hot Spots,'" *SAIS Review of International Affairs* 33, no. 2 (Summer–Fall 2013): 106.
67. Simon, "Safety and Security in the Malacca Straits," 29.
68. Bradford, "Shifting the Tides," 476, 480–81.
69. Simon, "Safety and Security in the Malacca Straits," 33.
70. Miha Hribernik, "Southeast Asia's Piracy Headache," *Diplomat*, February 15, 2015, http://thediplomat.com/2015/02/southeast-asias-piracy-headache/.
71. Raymond, "Piracy and Armed Robbery," 41.

Notes to Pages 185–188

72. Ian Storey, "The Triborder Sea Area: Maritime Southeast Asia's Ungoverned Space," *Terrorism Monitor* 5, no. 19 (October 2007), http://www.jamestown.org/single/?tx_ttnews%5Btt_news%5D=4465#.Vq6GYTYrJE6.
73. Bradford, "Shifting the Tides," 488.
74. ICC IMB, *Piracy and Armed Robbery against Ships Report*, 20, 29, 32.
75. Hribernik, "Southeast Asia's Piracy Headache."
76. ICC IMB, *Piracy and Armed Robbery against Ships Report*, 9.
77. "Piracy in Asia Worsens in Q1 2015," *IHS Fairplay*, April 23, 2015, http://fairplay.ihs.com/article/17610/piracy-asia-worsens-q1-2015.
78. Based on analysis of numbers from the International Maritime Bureau's *Piracy and Armed Robbery against Ships* reports from 2005, 2010, 2015, and 2016 covering years from 1994 to 2016.
79. ICC IMB, *Piracy and Armed Robbery against Ships Report*, 29; and Nincic, "Trends in Modern Piracy," 108, 114.
80. Kemp, "Crime on the High Seas: The World's Most Pirated Waters."
81. Hribernik, "Southeast Asia's Piracy Headache."
82. Hastings, "Geographies of State Failure," 216.
83. Bradford, "Shifting the Tides," 480.
84. Estimates from Raymond, "Piracy and Armed Robbery," 36; and Sittnick, "State Responsibility and Maritime Terrorism," 754. See also Teo, "Target Malacca Straits," 550–51.
85. Huang, "Building Maritime Security in Southeast Asia," 91.
86. Quoted in Bradford, "Shifting the Tides," 486–87.
87. Simon, "Safety and Security in the Malacca Straits," 30–31.
88. Joseph Trevithick, "The Philippine Navy Is Rusting Away: Manila Has to Make up for Decades of Neglect," *War Is Boring* (blog), January 6, 2015, https://medium.com/war-is-boring/the-philippine-navy-is-rusting-away-461bedea738e#.9hkawevgb.
89. Simon, "Safety and Security in the Malacca Straits," 31.
90. Ahmad Amri, "Piracy in Southeast Asia: An Overview of International and Regional Efforts," *Cornell International Law Journal Online* 1 (2014): 132; and Simon, "Safety and Security in the Malacca Straits," 30–31, 37.
91. Huang, "Building Maritime Security in Southeast Asia," 91.
92. Emmers, "The Threat of Transnational Crime in Southeast Asia," 9; and Huang, "Building Maritime Security in Southeast Asia," 88, 93. As Huang writes, Japan's Ocean Peacekeeping proposal was diluted to meet regional sovereignty sensibilities and reconstituted in 2002 as ReCAAP, which entered force in September 2006. Sheldon Simon, meanwhile, notes that Malaysia (home to the IMB's Piracy Reporting Center) has objected to the establishment of ReCAAP's Information Sharing Center in Singapore as an unnecessary redundancy. Simon, "Safety and Security in the Malacca Straits," 37–38.

93. Emmers, "The Threat of Transnational Crime in Southeast Asia," 10.
94. Bradford, "Shifting the Tides," 483–84.
95. Huang, "Building Maritime Security in Southeast Asia," 93, 96.
96. Raymond, "Piracy and Armed Robbery," 35.
97. Song, "Security in the Strait of Malacca," 104.
98. Huang, "Building Maritime Security in Southeast Asia," 87–88.
99. Bateman, "The Future Maritime Security Environment in Asia," 75.
100. Senia Febrica, "Securing the Sulu-Sulawesi Seas from Maritime Terrorism: A Troublesome Cooperation?," *Perspectives on Terrorism* 8, no. 3 (June 2014): 65.
101. Kemp, "Crime on the High Seas."
102. Amri, "Piracy in Southeast Asia," 131.
103. Bradford, "Shifting the Tides," 482.
104. Raymond, "Piracy and Armed Robbery," 38; and Simon, "Safety and Security in the Malacca Straits," 35–36.
105. Simon, "Safety and Security in the Malacca Straits," 30; and Huang, "Building Maritime Security in Southeast Asia," 90.
106. Simon, "Safety and Security in the Malacca Straits," 40.
107. Bradford, "Shifting the Tides," 486.
108. Huang, "Building Maritime Security in Southeast Asia," 99.
109. Raymond, "Piracy and Armed Robbery," 33.
110. Roach, "Enhancing Maritime Security," 97.
111. Luft and Korin, "Terrorism Goes to Sea," 67.
112. Luft and Korin, 64.
113. Young and Valencia, "Conflation of Piracy and Terrorism in Southeast Asia," 275.
114. Bateman, "The Future Maritime Security Environment in Asia," 62.
115. Huang, "Building Maritime Security in Southeast Asia," 89.
116. Teo, "Target Malacca Straits," 542.
117. Song, "Security in the Strait of Malacca," 101.
118. Febrica, "Securing the Sulu-Sulawesi Seas," 64.
119. Luft and Korin, "Terrorism Goes to Sea," 63.
120. Chatterjee, "Non-Traditional Maritime Security Threats," 89.
121. Emmers, "The Threat of Transnational Crime," 4.
122. Luft and Korin, "Terrorism Goes to Sea," 62.
123. Bradford, "Southeast Asian Maritime Security," 7.
124. Bradford, "The Maritime Strategy of the United States," 189.
125. Huang, "Building Maritime Security in Southeast Asia," 90.
126. Teo, "Target Malacca Straits," 551.
127. Luft and Korin, "Terrorism Goes to Sea," 63.
128. Song, "Security in the Strait of Malacca," 101; and Bradford, "Southeast Asian Maritime Security," 7.

129. Teo, "Target Malacca Straits," 552.
130. Bradford, "The Maritime Strategy of the United States," 190.
131. Song, "Security in the Strait of Malacca," 101, 104.
132. Song, 104.
133. Song, 122.
134. Teo, "Target Malacca Straits," 545.
135. Sittnick, "State Responsibility and Maritime Terrorism," 752.
136. Emmers, "The Threat of Transnational Crime in Southeast Asia," 1–2, 4.
137. Febrica, "Securing the Sulu-Sulawesi Seas," 75.
138. Bradford, "Southeast Asian Maritime Security," 13.
139. Luft and Korin, "Terrorism Goes to Sea," 61.
140. Young and Valencia, "Conflation of Piracy and Terrorism in Southeast Asia," 270.
141. Song, "Security in the Strait of Malacca," 103.
142. Khalid, "With a Little Help from My Friends," 426–27.
143. Khalid, 427.
144. Chatterjee, "Non-Traditional Maritime Security Threats," 83.
145. Nincic, "Trends in Modern Piracy," 112.
146. Hoe Pei Shan, "Piracy in Asia on the Rise," *Straits Times*, May 11, 2015, http://www.asiaone.com/asia/piracy-asia-rise.
147. Bateman, "The Future Maritime Security Environment in Asia," 59.
148. Chatterjee, "Non-Traditional Maritime Security Threats," 85.
149. Kemp, "Crime on the High Seas."
150. Raymond, "Piracy and Armed Robbery," 39.
151. E. L. Dabova, "Non-Traditional Threats in the Border Areas: Terrorism, Piracy, Environmental Degradation in Southeast Asian Maritime Domain," ISPRS/IGU/ICA Joint Workshop on Borderlands Modelling and Understanding for Global Sustainability, December 5–6, 2013, International Archives of the Photogrammetry, Remote Sensing and Spatial Information Sciences, volume 40-4/W3, http://dx.doi.org/10.5194/isprsarchives-XL-4-W3-51-2013, p. 54.
152. Dabova, 54.
153. Nincic, "Trends in Modern Piracy," 113.
154. James Wilson and George Kelling, "Broken Windows: The Police and Neighborhood Safety," *Atlantic*, March 1, 1982, http://www.theatlantic.com/magazine/archive/1982/03/broken-windows/304465/; and Khalid, "With a Little Help from My Friends."
155. Huang, "Building Maritime Security in Southeast Asia," 100.
156. Bradford, "The Maritime Strategy of the United States," 197.
157. Bradford, 197–98.
158. Robert Kaplan, "Center Stage for the Twenty-First Century: Power Plays in the Indian Ocean," *Foreign Affairs* 88, no. 2 (March–April 2009): 31.

159. Quoted in Amy Sawitta Lefevre and Andrew R.C. Marshall, "Special Report: Traffickers Use Abductions, Prison Ships to Feed Asian Slave Trade," Reuters, October 22, 2014, http://www.reuters.com/article/us-thailand-trafficking-specialreport-idUSKCN0IB0A320141022.
160. Scott Cheney-Peters, "Joint Patrols and U.S. Coast Guard Capacity," *Asia Maritime Transparency Initiative* (CSIS blog), April 1, 2015, http://amti.csis.org/joint-patrols-and-u-s-coast-guard-capacity/.
161. Bateman, "The Future Maritime Security Environment in Asia," 75.
162. Bateman, 75.
163. Young and Valencia, "Conflation of Piracy and Terrorism in Southeast Asia," 280.
164. Dupont, "Transnational Crime, Drugs, and Security," 453.
165. Chris Trelawny, "Maritime Security Beyond Military Operations," *RUSI Journal* 158, no. 1 (2013): 50.
166. Simon, "Safety and Security in the Malacca Straits," 41.
167. Nincic, "Trends in Modern Piracy," 105.
168. Chatterjee, "Non-Traditional Maritime Security Threats," 80.
169. Bradford, "Southeast Asian Maritime Security," 13; see also Song, "Security in the Strait of Malacca," 103.
170. Kaplan, "Center Stage for the Twenty-First Century," 19.
171. Quoted in Kaplan.
172. Simon, "Safety and Security in the Malacca Straits," 39.
173. Chatterjee, "Non-Traditional Maritime Security Threats," 78.
174. Dabova, "Non-Traditional Threats in the Border Areas," 55.
175. Chatterjee, "Non-Traditional Maritime Security Threats," 78.
176. Nincic, "Trends in Modern Piracy," 106.
177. Quoted in Cheney-Peters, "Joint Patrols and U.S. Coast Guard Capacity."

Chapter Eight. Charting a Course
1. Michael White, Henry Fradella, and James Coldren Jr., "[Updated] Why Police (and Communities) Need 'Broken Windows,'" *Crime Report*, August 11, 2015, http://www.thecrimereport.org/viewpoints/2015-08-why-police-and-communities-need-broken-windows; and George Kelling, "Don't Blame My 'Broken Windows' Theory for Poor Policing," August 11, 2015, http://www.politico.com/magazine/story/2015/08/broken-windows-theory-poor-policing-ferguson-kelling-121268?o=0.
2. Kelling, "Don't Blame My 'Broken Windows' Theory."
3. Joshua Tallis, "Other Than War: HA/DR and Geopolitics," Center for International Maritime Security (CIMSEC) *NextWar* (blog), March 28, 2016, http://cimsec.org/war-hadr-geopolitics/23591.
4. Chris Trelawny, "Maritime Security Beyond Military Operations," *RUSI Journal* 158, no. 1 (2013): 51.
5. Trelawny, 51.

Index

Page numbers in italics refer to photographs.

Abu Nidal Organization, 114
Abu Sayyaf Group, 193, 194
ACCP. *See* Association of Caribbean Commissioners of Police
acquisitions vs. strategy, 31–34
Afghanistan, Navy's contribution to inland operations in, 3
Africa, demographic shifts in, 15. *See also specific countries*
Africa Command (AFRICOM), 143, 148, 168–70
Africa Partnership Station (APS), 168–70
African Maritime Law Enforcement Partnership (AMLEP), 169–70
African Pride (pirate tanker), 153
African Union: *2050 Africa's Integrated Maritime Strategy* (2012), 28
AFRICOM. *See* Africa Command
Ai Maru (tanker), 190
AIDS. *See* HIV/AIDS
Air Corps and bureaucracy, 22
al-Faruq, Omar, 194
Alfonzo, Don (pseud.), 89–90
Alliance Maritime Strategy (NATO, 2011), 28
Al-Qaeda, 147, 151, 192–96
Amaechi, Chibuike Rotimi, 167
AMLEP. *See* African Maritime Law Enforcement Partnership
amphetamines, 178
Amri, Ahmad, 190
Anderson, Stephen, 145
Anegada Passage, radar installations in, 71
Angola, 143, 164

APS. *See* Africa Partnership Station
Arellano Felix organization, 71, 86
Arias, Martín, 90
arms trafficking, 91–96, 151, 155
Arthur, Owen, 67–68
Association of Caribbean Commissioners of Police (ACCP), 94, 95
Association of Southeast Asian Nations (ASEAN), 177–78, 182, 184, 188, 190
AUC. *See* United Self-Defense Forces of Colombia

Badawi, Abdullah Ahmad, 189
Bahamas: community policing in, 94; drug trade in, 74–76; migration in, 104, 106
Bailey, John, 73
Bakassi Boys, 154–55
balloon effect in drug enforcement operations, 71, 82
Barakat, Assad Ahmad, 112
Barbados: drug trade in, 74, 75; Regional Police Training Centre, 94
Bateman, Sam, 177–78, 189, 193, 197, 199
Bay Islands, drug trade in, 81
Beith, Malcolm, 73
Belize: community policing initiatives in, 94; drug trade in, 74, 75, 79, 81, 91
Benin, piracy in, 164, 169
Biekro, Geoffrey, 144
Bitzinger, Richard, 202
blood diamonds, 151–52
Blue Atlantic (ship), 145
blue waters. *See* naval strategy

253

Boko Haram, 147, 167
Bolivia, coca cultivation in, 70
Bonga (offshore oil platform), 154
Bouterse, Dino, 124
Boyce, Ralph, 189
Bradford, John: on counterpiracy operations, 180, 182, 184, 185, 187, 191; on counterterrorism operations, 194; on human trafficking, 179; on humanitarian relief operations, 198; on ReCAAP, 182, 190; on Southeast Asia's reliance on maritime environment, 176; on transnational crime, 196
Brantley, Mark, 111
Bratton, William, 42–49, 94, 127–28
Brazil, 20
Breen-Smyth, Marie, 58
broken windows theory, 37–43; Caribbean application of, 60–61, 65, 69, 95–96, 101, 103, 108, 118, 127, 136–37, 209; Coast Guard and, 127–31; on context and environmental influences, 7–8, 19–20, 35, 37–41, 46, 50–53, 56–59, 61, 65–67; on disorder, 39–44, 47–48, 50–55, 58–59; on Gulf of Guinea, 149, 150, 159, 162, 172–73; hearts and minds vs., 19; index to measure, 222n62; maritime security and, 6–9, 25, 35, 45, 141; New York City crime and, 46–49, 55; New York City subway and, 43–46; public health and, 50–51, 101; racism and classism challenges to, 41–43, 58, 205–6; social norms in, 57–58; on Straits of Malacca and Singapore, 175, 177, 180, 192–203; technologies of, 94
broken Yankees hypothesis, 54, 55
Bronx, New York City, 38–39
Brown, Ben, 58
brown waters. *See* littoral areas
budgets and sequestration: for community policing, 129; counternarcotics and, 82–83, 117; human trafficking and, 108; littoral combat ships and, 32–33; in Straits of Malacca, 188;

terrorism and, 121; theory's influence on, 22–23
Bueger, Christian, 4
bureaucracy: disaggregation of operations and, 25, 68–69, 128; influence on theory, 21–22; institutional change and, 48; international partnerships and, 132; law enforcement and, 41; network culture in, 127–28; Sinaloa Cartel and, 73
Bush, George W., 19

Calabar River, 163
Cali Cartel, 70, 86
Cameroon, piracy in, 164, 169
car experiment on environmental influences, 38–39
Cardenas Guillen organization, 71
Caribbean Basin: broken windows theory on, 60–61, 65, 69, 95–96, 101, 103, 108, 118, 127, 136–37, 209; Coast Guard policing in, 8, 68, 83, 127, 129–30; community policing programs in, 94–95; counternarcotics in, 81–85; crime and homicide rates in, 89–93; demographic shifts in, 15; domestic migration in, 100, 102, 104, 107–8, 110; drug culture in, 89–92, 133–34; gang presence in, 92–93, 103, 106; geography of, 69, 132; human trafficking in, 96–102; hybrid threats to, 67–68; *International Narcotics Control Strategy Report* on, 74, 93; irregular migration in, 102–7; maritime security initiatives in, 85–86, 95–96, 136–37; misgovernance in, 122–26; money laundering in, 118–22; police initiatives in, 93–96, 108, 124, 129–32; terrorism and, 109–15, 117; tourism and, 67, 100, 107–8; unemployment in, 103, 104. *See also specific countries*
Caribbean Basin Security Initiative, 93
Caribbean Community (CARICOM), 95, 107, 109, 132

Index

Carillo-Fuentes organization, 71
cartels. *See* nonstate actors and transnational crime organizations; *specific cartels*
Casteel, Steven, 115, 120
Cayman Islands and money laundering, 119
CBP. *See* Customs and Border Protection
Center of Excellence for Stability Policing Units (NATO), 18
Central America. *See specific countries*
Centre for Criminology and Criminal Justice (University of the West Indies), 94
Cepeda Ulloa, Fernando, 135
Chacón, David, 90
Chase (WHEC 721), 148–49
Chatterjee, Aditi, 178–80, 193–94, 196, 197, 201
Cheney-Peters, Scott, 198–99
children as trafficking victims, 96, 97, 99–101, 156, 230n51
China: ASEAN opposed by, 188; irregular migration and, 102; maritime role in Southeast Asia, 176, 179; PLA Navy, 32, 169; West Africa and, 169, 170
cities: "feral," 15–17; maritime space compared to, 60–61; police operations in, 20; population growth and crime in, 15. *See also specific cities*
citizenship and visas, 110–11
Clean Car Program (New York City), 43
climate change: *CS21* on, 31; pressures induced by, 15, 26, 34, 99, 105–6; in Southeast Asia, 176
Clinton, Bill, 19, 71
Coast Guard: Africa Partnership Station and, 170; broken windows theory and, 127–31; Caribbean Basin policing by, 8, 68, 83, 127, 129–30; counternarcotics and, 83–87; *CS21* and, 30; migrant interdiction by, 103; mutual legal assistance treaties and, 85; Nigerian Navy and, 158; Operation Martillo and, *64*, 82–83; policing techniques of, 8; Southeast Asia and, 183, 198–99; *Western Hemisphere Strategy* (2014), 126–29, 133
Coastal Riverine Force, 30
cocaine: consumption of, 55, 56; geonarcotics of, 70–81; health care and enforcement costs of, 86; narco-submarines and, 76–77; value of, 70–72, 86–87; West Africa and, 144–45
Coke, Christopher, 122–24
Cole, Patrick Dele, 153
Coles, Catherine, 38, 40–43, 45, 47, 49
Colombia: drug trade in, 70–74, 76, 78, 81, 86, 91, 113–14; FARC (Revolutionary Armed Forces of Colombia), 113–15, 148; human trafficking in, 98–100; terrorism and paramilitary groups in, 113–14
Columbia-class submarines, 31
Comfort (T-AH 20), 108
commerce and shipping: through the Caribbean, 69; ease of, 75; in naval strategy, 14, 24–25, 27; through the Straits of Malacca and Singapore, 176–77, 179–80, 182, 245–46n6
community policing theories, 6, 18, 20, 35, 42, 66, 94–96, 168
COMPSTAT (comparative statistics) system, 94
context and environmental influences: in broken windows theory, 7, 8, 19–20, 35, 37–41, 46, 50–53, 56–59, 61, 65–67; Caribbean Basin and, 134–37; causative relationship to crime, 52–53, 61; in Gulf of Guinea, 149–50; human trafficking and, 101–2; irregular migration and, 103; misgovernance and, 125; perception of, 57–58, 66, 118, 125, 135–36, 149–50, 171, 209, 214; students' responses to, 51–52, 118
Convention for the Suppression of Unlawful Acts against the Safety of Maritime Navigation (1988), 29, 188
Cooperation and Readiness Afloat, 183, 199

Cooperative Strategy for 21st Century Seapower, A (*CS21*, 2007), 8, 27, 28, 30–34, 212, 213, 218n61
Cope, John, 111, 112
Corbett, Julain, 2, 6, 24
Coronado (LCS 4), *174*
corruption. *See* misgovernance; money laundering
Costa Rica: community policing initiatives in, 94; drug trade in, 81, 90; human trafficking in, 98, 100; Ministry of Public Security, 94
Côte d'Ivoire, piracy in, 164
counterinsurgency theories, 19
counternarcotics: Caribbean deployment of, 81–85; counterterrorism vs., 68–69, 115; expense and efficacy of, 86–87; funding for, 82–83; geography of, 70–71; Gulf of Guinea and, 148; human trafficking enforcement compared to, 98; *INCSR* on, 74; international partnerships in, 132; multidimensionality of, 65; routes and trends impacted by, 71, 73–77
counterpiracy, 160–61, 166–72, 175, 181–85, 190, 191, 197, 202–3
counterterrorism: counternarcotics vs., 68–69, 115; in Southeast Asia, 176, 183; U.S. priority on, 191
coyote, 102
Crawford, Adam, 58
crime: broken windows theory on, 6–9, 37–41, 47–48, 52–53, 57–59, 61, 65; Caribbean context for, 89–96, 101–2; Gulf of Guinea's context for, 144, 150–59; local vs. transnational, 89, 91, 93–94, 96; multidimensionality and, 44, 56, 66, 104–5, 108, 117–18, 167, 208; in New York City, 47–49, 55, 127–28; New York City subway and, 43; population growth and, 15; rates of (1970s–1990s), 40, 47–48, 56; in Southeast Asia, 195–202; trade and migration and, 17. *See also* drug trade; pirates and piracy; trafficking
Cuba and migration, 103, 104, 108

Customs and Border Protection (CBP), 73, 83, 84, 86

Dabova, E. L., 197
Dagorn, René-Eric, 172
Daley, Matthew, 195
Davies, Omar, 91
DEA. *See* Drug Enforcement Administration
debt bondage. *See* human trafficking
Defense Department: on drug trade revenue, 72; "Trends and Shocks" project (2006), 15
demographic shifts, 14–16, 18, 26, 34, 35, 207. *See also* migration
Deosaran, Ramesh, 95
deportation, 106–7
Design for Maintaining Maritime Superiority, A (2016), 31, 33
Dewi Madrim (tanker), 193
diamond trade and blood diamonds, 151–52
disaster relief. *See* humanitarian assistance and disaster relief
disorder: in broken windows theory, 39–44, 47–48, 50–55, 58–59; in Caribbean Basin, 117–18, 125; Skogan on, 220n9; in Straits of Malacca and Singapore, 187, 199. *See also* context and environmental influences
Doe, Samuel, 147
Dominica, investor visas and, 111
Dominican Republic: community policing initiatives in, 94; drug trade in, 74, 75, 78–79, 87; human trafficking in, 99, 100; migration and, 78–79, 100, 103, 104, 106; remittance networks and, 78; security initiatives in, 93
Drug Enforcement Administration (DEA), 74, 75, 81, 86, 90
drug trade: adaptation to enforcement, 76–77, 81, 82, 86–87; aircraft in, 70–71, 75; Caribbean routes for, 73–78; cost of interdiction in, 86; crime and homicide rates tied to, 55–56, 90–96; development of,

69–73; dominance over legitimate enterprise, 72–73, 89–90; in Gulf of Guinea and West Africa, 144–50; Hezbollah and, 112; human trafficking tied to, 101–2, 109–10; *International Narcotics Control Strategy Report* (*INCSR*) on, 74; misgovernance and, 122–26; money laundering and, 118–22; multidimensionality of, 133–36; payment in kind in, 71, 80, 146; public safety concerns and, 79–80; terrorist organizations and, 109–15, 120–21. *See also* counternarcotics; *specific drugs*
Dupont, Alan, 178, 200

Ebola, 105, 157
ecstasy, 74, 81
Ecuador, human trafficking in, 98
Eisenhower, Dwight, 23
El Salvador: community policing initiatives in, 94; drug trade in, 81, 91, 92; human trafficking in, 98; migration from, 102
Elhakim, Dedy Fauzi, 177
Elkus, Adam, 15, 17, 20
Ellis, Stephen, 145–50
ELN. *See* National Liberation Army
Emmers, Ralf, 188
Enforcement, Operation (1989), 44
environmental influences. *See* context and environmental influences
Equatorial Guinea, pirate attack against, 163–64
Escobar Gaviria, Pablo, 70, 86
European Union: counterpiracy initiatives of, 169, 172; as drug market, 144; *Maritime Security Strategy* (2014), 28–29
Europol, 84–85
exclusive economic zone, 4
extortion. *See* fare evasion and extortion
Eyadéma, Gnassingbé, 147
Eyes in the Sky program, 182, 190

falcons (lookouts), 90
Farah, Doug, 151

FARC. *See* Revolutionary Armed Forces of Colombia
fare evasion and extortion, 43, 44, 46, 47, 67
Fargo, Thomas, 189
Febrica, Senia, 193, 196
Federal Bureau of Investigation, 68, 86
Federal Deposit Insurance Corporation, 121
Felíx criminal organization, 86
feral cities and regions, 15–17
financial interdictions in drug trade, 83–84
fish and fishing industry: crimes reported by, 187–88; drug trade and, 79–80, 90, 134, 135, 145; forced labor and, 156; illicit trade of, 151–53, 156, 166, 178–79; piracy and, 166, 172; role of, 240n70; terrorism and, 194
flag of convenience laws, 76
Fleet Week New York (2017), *36*
Florida, drug trade through, 73, 77
Florida Strait, 69
forced labor. *See* human trafficking
foreign policy, 13, 22
Fort McHenry (LSD 43), 170
Forum, Jamaica, 79–80
Free Aceh Movement, 193
Freedom (LCS 1), *12*
French Coast Guard, 85
Frick, Matthew, 16
fully submersible vessels (FSVs), 77

Gallatin (WHEC 721), 149
gangs and gang violence, 92, 103, 106
Garden Parish, Operation (2011), 123–24
Garner, Randy, 198
geonarcotics, 69, 70
Gerald Ford–class carriers, 31
Germond, Basil, 4, 5, 28
Ghana: counternarcotic efforts in, 145; gun sales in, 151; human trafficking in, 155, 156
Gilpin, Raymond, 143
Gladwell, Malcolm: *The Tipping Point*, 37–38, 41, 46, 48, 56, 66, 118
Global Trafficking in Persons Report (2012), 97, 98, 101

go-fast boats, 70, 75–77, 80–81
gonorrhea spread, 50–51, 54
Gosse, Philip, 163
Gottlieb, Yaron, 171
graffiti, 7, 37, 41, 43–44, 51
Grant, Audra, 146, 151
Greater Antilles, drug trade in, 75
green waters. *See* littoral areas
Greenert, Jonathan, 31
Gregorio del Pilar (Philippines), 183
Grenada, citizenship program in, 111
Grenadines, drug trade in, 76
Griffin, Clifford, 95
Griffith, Ivelaw: *Caribbean Security in the Age of Terror*, 67, 96
Guatemala: community policing initiatives in, 94; corruption in, 135; drug trade in, 81, 91, 92; human trafficking in, 98, 99; migration from, 102
Guinea, piracy in, 164
Guinea-Bissau, misgovernance in, 147–48
Gulf of Guinea and West Africa: broken windows theory on, 149, 150, 159, 162, 172–73; counterpiracy initiatives in, 168–72; drug trade in, 144–50; geography and demography of, 142–44; irregular migration in, 155–57; Latin American cartels in, 145–46, 148; militarization in, 168; misgovernance in, 157–59; money laundering in, 152–53; oil trade in, 143–44, 153–55; piracy in, 160–67; strategic context for, 8–9, 149; vigilantism in, 154–55
Gunn, David, 43
guns and gun-related crimes, 91–96, 126, 151, 153, 155
Gurita, Operation (2005), 187
Guzman, Chapo, 71–73, 134

HA/DR. *See* humanitarian assistance and disaster relief
Hagel, Chuck, 32
Haiti: drug trade in, 75; earthquake in (2010), 99, 105, 130; human trafficking in, 99, 100; migration from, 103, 104, 106
Hamilton (WHEC 715), 183
Harcourt, Bernard, 54
Hastings, Justin, 177, 180
Haynes, Peter, 24, 29, 30; *Toward a New Maritime Strategy*, 22
hazardous waste, smuggling of, 197
health. *See* public health
hearts and minds, 19
heroin, 146, 147
Hezbollah, 16–17, 110–12, 114–15
Hispaniola, 78, 79, 103, 106
HIV/AIDS, 101, 105, 157
Hoffman, Frank, 17–18
Homeland Security, Department of, 120, 121
homelessness and the homeless, 41–42, 44, 99
homicide rates in Caribbean, 91–96
Honduras: drug trade in, 80–82; homicide rates in, 91–93; human trafficking in, 98, 100; migration from, 102
Huang, Victor, 187–89, 190, 193, 194, 198, 249n92
Hulse, Janie, 111, 112
human rights, 108
human trafficking, 96–102, 106–8, 110, 126, 152, 155–57, 179, 230n51
humanitarian assistance and disaster relief (HA/DR), 31, 33, 130, 176, 198
Hurricane Maria (2017), 79
Hutchinson, Asa, 115, 120
Hutchinson, Steven, 58
hybrid threats: broken windows theory on, 207; in Caribbean, 67–68, 77, 87, 133; Coast Guard responses to, 128; in Gulf of Guinea, 161–62; narco-subs as, 77; Navy responses to, 34, 35; piracy and, 193, 196; strategic responses to, 14, 17–18, 20–21; white papers on, 25–26. *See also* multidimensionality

ICE. *See* Immigration and Customs Enforcement

ICG. *See* International Crisis Group
IDF. *See* Israel Defense Force
ILO. *See* International Labour Organization
IMB. *See* International Maritime Bureau
Immigration and Customs Enforcement (ICE), 79, 86
INCSR. *See International Narcotics Control Strategy Report*
Indonesia: corruption in, 187; counter-piracy initiatives in, 182, 183, 187; geopolitics of, 177, 179, 188–89; illicit fishing in, 178–79; Ministry of Home Affairs, 182; piracy and, 181, 184–87, 190; terrorism and, 193, 194; tsunami (2004) in, 184, 198
Innes, Martin, 59
institutional decapitation, 86, 87
International Crisis Group (ICG), 143, 170–72
International Labour Organization (ILO), 97, 99, 152, 156
International Maritime Bureau (IMB), 160, 163, 180, 181, 184–86, 190, 247n31
International Maritime Organization, 169, 181, 184, 214
International Narcotics Control Strategy Report (INCSR): on citizen safety, 93; on Coast Guard budget, 83; on Coast Guard law enforcement detachments, 84; on corruption, 135; on drug-producing countries, 74; on drug-transit countries, 74; on gang networks, 106; on Jamaica, 80; on money-laundering, 119, 121; on West Africa, 145, 146, 149, 150
International Organization for Migration (IOM), 97, 100, 104, 106, 124
international partnerships in law enforcement: in Caribbean Basin, 84–85, 126–27; in Gulf of Guinea, 168–72; Italian and Australian role in, 217n24; in Southeast Asia, 177–78, 181–84, 187–92; threat-specific approach, 85
Interpol, 84–85, 169

interstate war vs. nontraditional conflicts, 18–20, 23–25, 34–35
investor visas, 110–11
involuntary servitude. *See* human trafficking
IOM. *See* International Organization for Migration
IRA. *See* Provisional Irish Republican Army
Iran, Caribbean connections of, 111, 112, 114–15
Israel Defense Force (IDF), 20
IUU fishing. *See* fish and fishing industry
ivory, 153

Jamaica: corruption in, 135; drug trade in, 72, 74, 75, 79–80; garrison districts in, 16, 78, 94, 122–24; homicide rates in, 91–92; migration and, 104; militarization of police in, 93–94; public safety concerns in, 79–80; remittance networks in, 78
Jang, Sung Joon, 53
Japan: Coast Guard, 182–83; counter-piracy initiatives of, 182–83, 188, 191, 249n92; oil imports and, 176
Jemaah Islamiyah (JI), 194
JIATF-South. *See* Joint Interagency Task Force South
Johannesburg, as feral city, 16
John C. Stennis (CVN 74), *12*
Johns Hopkins University study on neighborhood relationships, 50
Johnson, Byron, 53
Johnson, Lyndon, 19
Joint Interagency Task Force South (JIATF-South), 77, 81–84, 87
Joint Riverine Training Teams, 83

Kalifornia (Indonesian ferry), 194
Kamal-Deen, Ali, 142, 154, 159, 161–65, 167
Kaplan, Robert, 198, 201
Karachi, as feral city, 16
Kearsarge (LHD 3), *36*
Kelling, George, 38–43, 45–50, 53, 118, 205, 206

Kelly, John: on Caribbean trafficking infrastructure, 109, 133; on corruption, 135; on drug trade, 72, 77, 78; on funding for counternarcotics, 117; on human trafficking, 105, 108; on local crime, 93, 96; on migration, 102–3; on militarization of police, 134; on MS-13, 106; on multidimensionality of criminal networks, 105; on Navy/law enforcement collaboration, 84, 85; on Operation Martillo, 82–83; on resources for humanitarian assistance, 108; on Southern Command's budget, 82–83; on transnational crime, 133

Kelly, Raymond, 46

Khalid, Nazery, 183, 196, 198

kidnappings and ransoms, 100

Kilcullen, David: on community-based policing, 20; on competitive control, 16; on connectedness, 17; on demographic shifts, 21; on irregular operations, 19; *Out of the Mountains*, 14; on police command efficiency, 48; on police militarization, 94; on remittance networks, 78; on Tivoli Gardens assault, 122–23

Kiley, Robert, 43

Kingston, Jamaica, 16, 92, 122–24, 162

Knox, Dudley, 2–3

Korin, Anne, 193, 196

Krulak, Victor "Brute," 49

Kuber (fishing trawler), 1

Kufuor, John, 168

Latin America: demographic shifts in, 15. *See also* Caribbean Basin; *specific countries*

law enforcement. *See* international partnerships in law enforcement; police and law enforcement

LCSs. *See* littoral combat ships

Le Sage, Andre, 147, 148, 151, 154, 155, 159, 163

Lebanon, Hezbollah in, 16–17

Lehr, Peter, 193

Lekir (Malaysia), *174*

Lekiu (Malaysia), *174*

Leonhart, Michele, 125

Lesser Antilles, drug trade in, 71, 75

Liberia: drug trade in, 147; gun trade in, 151; naval strength of, 143

littoral areas: connectedness through, 18, 21, 31; defined, 3–4; demographic shifts in, 14–15, 29, 45, 176, 207; environmental influences in, 41; hybrid threats from, 17–21, 24–25, 143–44, 150–51, 161–62, 213–14; littoral combat ships and, 32–33, 198; Navy strategy and, 30–35, 130; strategic challenges of, 1–4, 13, 20–21, 211. *See also specific littorals*

littoral combat ships (LCSs), *12*, 32–33, 198

Los Zetas, 101, 115

Ludwig, Jens, 54

Luft, Gal, 193, 196

Mahan, Alfred Thayer, 22–24, 213; *The Influence of Sea Power upon History*, 23

Mahnken, Thomas, 217n24

Maingot, Anthony, 78–79, 87, 125, 130–31

Malacca Straits. *See* Straits of Malacca and Singapore

Malaysia: drug trade in, 178; geopolitics of, 177, 179, 188–89, 249n92; Maritime Law Enforcement Agency, 182–83, 190, 194; piracy and, 181, 185–86, 190; terrorism and, 193, 194

Malsindo, Operation (2008), 182

Mara Salvatrucha (MS-13), 106

maras (gangs), 92

marijuana, 70, 75, 79, 80, 91

Marine Corps: *CS21* and, 30; Forces South, 83; *Marine Corps Operations* manual, 30

Maritime Law Enforcement Initiative, 183–84

maritime security: acquisitions vs. strategy, 31–34; broken windows theory and, 6–9, 35, 41, 59–61, 65,

67, 129, 136–37, 199–203, 206–9; Caribbean initiatives for, 85–86, 95–96, 136–37; Coast Guard and, 126–31; counternarcotics and, 81–82, 85–87; defined, 4–6, 216n16; feral cities and, 16; foreign policy and, 13; global/local approach to, 31, 89, 136, 171, 213; littoral combat ships and, 33; littoral stability and, 20–21; multidimensionality of, 25, 109, 131–32, 195–202, 208–9; Navy literature on, 27–28, 30–34; sea power vs., 23–25, 59; sovereignty concerns and, 143, 160, 177–79, 188–91, 241n94; terrorism and, 109; white papers on, 26–27. *See also* naval strategy
Market Time, Operation (1965), 126
Martillo, Operation (2012), *64*, 82–83
McFadden, David, 110
McKenzie, Desmond, 92
Medellín Cartel, 70, 86
MEND. *See* Movement for Emancipation of the Niger Delta
Mercy (T-AH 19), 198
methamphetamine, 74, 81, 115
Mexico: cartels in, 71–73, 106–7, 115, 144; corruption in, 135; drug trade in, 71–74, 80–81, 86–87; human trafficking in, 98, 100, 101; migration and, 102, 103
Mexico City, as feral city, 16
Michel, Charles, 77, 80, 83, 84, 109, 117, 128
migration: Caribbean culture of, 100; climate change and, 105–6, 179; *CS21* on, 31; domestic, in Caribbean Basin, 100, 102, 104, 107–8, 110; environmental degradation and, 100, 157; feral cities and, 17; irregular, 98, 102–8, 117, 155–57, 179; maritime deaths from, 104; repatriation and, 106
military urbanism, 20
Millett, Richard, 69–70
Mischel, Walter, 55, 57
misdemeanors, 41, 44, 47

misgovernance, 122–26, 147–48, 154–55, 157–59, 171–72, 187
Mogadishu, 15, 94
Mona Passage: Haitian migrants in, 103; maritime traffic in, 69; radar installations in, 71
money laundering, 8, 118–22, 152–53, 155
Moro Islamic Liberation Front, 193, 194
Moro National Liberation Front, 193
Movement for Emancipation of the Niger Delta (MEND), 153–55, 158, 162, 165
MS-13. *See* Mara Salvatrucha
MTA. *See* New York City Metropolitan Transit Authority
Mullen, Michael, 25, 30
multidimensionality: Caribbean Basin and, 117–18, 131–34, 137; counternarcotics and, 65, 81–82; of crime, 44, 56, 66, 104–5, 108, 117–18, 167, 208; Gulf of Guinea and, 144, 147, 149, 155, 159, 167; of money laundering, 121; of piracy, 192, 196–202; Straits of Malacca and Singapore and, 185; strategic responses across, 25, 108, 127–28
Mumbai, Pakistani terrorist attack in (2008), 1–2, 215n2
Murphy, Martin, 153–55, 157–58, 161, 162, 172
Murray, Rodwell, 124
mutual legal assistance treaties, 85

Nagl, John, 33; *Learning to Eat Soup with a Knife*, 48
Naím, Moisés: on border impacts on counternarcotics operations, 85–87; on crime in Lagos, 151; on FARC, 113, 114; on gun crime in Latin America, 91; on Guzman, 134; on human trafficking, 98, 100–101; on irregular migration, 102; on narcotics trafficking operations, 89
narco-submarines, 76–77, 81, 136, 146
narcotics. *See* drug trade; *specific drugs*

Nataly (narcotics trafficking ship), 75
National Defense Strategy (*NDS*, 2008), 25, 26
National Liberation Army (ELN), 114
National Military Strategy (*NMS*, 2011), 25–26
National Security Strategy (2010), 18, 25–26
National Strategy for Maritime Security (*NSMS*, 2005), 26
National Union for the Total Independence of Angola, 151
NATO. *See* North Atlantic Treaty Organization
natural gas, 246n6
Naval Doctrinal Publication 1, 27
Naval Forces Southern Command, 83
Naval Operations Concept 2010 (*NOC*), 27
naval strategy: broken windows theory's relevance to, 49, 130, 212–14; bureaucracy's role in, 21–23, 206–7; Coastal Riverine Force, 30; *CS21* and, 30–34, 212; doctrine vs., 27; Gulf of Guinea and, 169–70; littorals in, 1, 3, 6, 13, 21, 29–35, 130; for maritime security, 5–7, 9, 13, 23, 31–35, 84; in peacetime, 211–12; police operations and, 18, 87; postmodern, 24–25, 30–32; recommendations for, 213–14; sea power and, 21–25, 59, 212; Southeast Asia and, 183; for unconventional threats, 21, 25; white papers on, 25–26
Nevis. *See* St. Kitts and Nevis
New Orleans, public health in, 50–51
New York City crime rates, 47–49, 55, 127–28, 208
New York City Metropolitan Transit Authority (MTA), 43, 44
New York City subway, 43–46, 208
New York Police Department (NYPD), 40, 42, 66, 94
Nicaragua: drug trade in, 81; human trafficking in, 98
Nigeria: corruption in, 150; drug trade in, 146; geography and demography of, 142–43; human trafficking in, 155–56; Navy, 149, 152, 153, 158; oil exportation from, 143–44, 147, 154; oil theft in, 153–55, 163; piracy in, 160–66, 169, 171, 186; remittance networks and, 146; sovereignty concerns and, 241n94; state weakness of, 158–59, 171; vigilantism in, 154–55
Nincic, Donna, 196–98, 200, 202
NMS. See National Military Strategy
NOC. See Naval Operations Concept 2010
nonstate actors and transnational crime organizations: cash smuggling and, 120–21; Coast Guard's strategy on, 126–27, 130–31; connectedness and, 17, 106–7, 109, 115; deportation and, 106–7; diversification and, 76, 81, 87, 101, 127, 171, 197; feral cities empowering, 16–17, 125; hybrid threats from, 18, 77, 133; maritime security and, 5, 6, 14; navies' role in threats from, 211–12; organizational structure of, 73, 86, 87, 89–91, 101, 104–5, 127, 130–31, 146, 197; piracy and, 165–67, 171, 193–94, 196–97; terminology of, 224n15; white papers on, 26, 34. *See also specific cartels*
Noriega, Roger Francisco, 112
North Atlantic Treaty Organization (NATO), 18, 19, 28, 169, 170, 207
Norton, Richard, 15
NSMS. See National Strategy for Maritime Security
NYPD. *See* New York Police Department

Obama, Barack, 19, 149
Obuchi, Keizō, 188
oil: corruption and, 158; exports, 143–44, 147, 245–46n6; Singapore and, 177; spills and pollution, 152, 154; theft of, 153–55, 163, 165–67
Okpabana (Nigeria), 149
Onuoha, Freedom, 142, 143, 151, 160
operational domains, 3

order maintenance: legislation for, 42; in maritime security, 59; military's response to, 41, 42; New York City crime rates and, 47–48; New York City subway and, 44, 45; nonstate actors providing, 16–17, 34; poverty vs., 44, 45, 47–49, 50
Orinoco River, 76
Osinowo, Adeniyi Adejimi, 142, 152, 165, 172–73
Otto, Lisa, 152–54, 160–61, 165, 171
Our Lady Mediatrix (Philippine ferry), 194

Pacific Partnership initiative, 198
Palo Alto, California, 38–39
Panama: drug trade in, 81; human trafficking in, 99; National Border Service, 94–95; National Police, 94
Panama Canal, 69
pandillas (youth gangs), 92
Parnell, Isaiah, 123
Patterns of Global Terrorism 2002 (State Department), 68, 112
Perl, Raphael, 115
Peru, coca cultivation in, 70
petty crime, 7
Philippine Navy, 187–88
Phillips, Richard, 197
pirates and piracy: broken windows theory on, 141–42, 159, 162, 163, 167–71, 200–201, 210; costs induced by, 180; *CS21* on, 31–32; defined, 160–61, 247n31; in feral cities, 16, 17; in Gulf of Guinea, 160–67; incidence of, 162–64, 180, 181, 184, 186–87; initiatives against, 168–71, 181–85; multidimensionality of, 192, 196–202; politics and, 161–62, 196; in Straits of Malacca and Singapore, 179–92, 197; strategic context for, 8–9; tactics of, 164–67, 180; terrorism and, 166–67, 192–96. *See also* counterpiracy
Ploch, Lauren, 168, 170
PMSCs. *See* private maritime security companies

police and law enforcement: in Caribbean Basin, 93–96, 108, 124, 129–32; corruption and, 124–26; counternarcotics and, 65, 70–71, 77–78; counterpiracy and, 168–72; environmental disorder and, 7, 38–41, 58–59, 66; everyday security and, 58–59; illicit fishing and, 179; institutional structure in, 25, 41, 68–69, 127–28; LEDETs and, 129–30; maritime capabilities of, 162; militarization of, 93–94, 134, 168, 205–6; in New York City, 43–49; Posse Comitatus Act (1878) and, 84; racism and classism challenges to, 41–43, 58, 205–6; statistical software in, 94; traditional model of, 219n5; in West Africa, 154–55, 168–72. *See also* broken windows theory; international partnerships in law enforcement; prevention vs. crime fighting
police vs. military operations: Coast Guard and, 126–27; convergence of, 45, 48–49, 93–94; *CS21* on, 30; Mahan and Corbett on, 24; against nonstate actors, 17–21, 122–23, 167–68, 199–200; stability from balance of, 20–21, 35, 137, 162, 213
polleros (professional smugglers), 102
poppy cultivation, 178
population growth. *See* demographic shifts
Port Harcourt–Nembe, 163
Posse Comitatus Act (1878), 84
poverty: gonorrhea spread and, 50–51; human trafficking and, 99; in littoral terrain, 1; urbanization and, 15, 16, 34
precursor chemicals, 74, 81, 115, 121, 145
prevention vs. crime fighting, 30, 40–42, 61, 67, 85–86, 93–94, 127, 133, 136
Princeton (CG 59), *204*
private maritime security companies (PMSCs), 158–59, 241n94

prostitution, 96, 100
Provisional Irish Republican Army (IRA), 114
public health, 50–51, 53, 54, 101, 105, 108, 156–57, 179
Puerto Rico: drug trade in, 70–71, 73–75, 78–80, 92; homicide rates in, 92; interdiction efforts in, 83; migration and, 103

Quadrennial Defense Review (QDR), 14, 17, 25, 34
quality-of-life policing, 39–41, 47, 66, 199

racism and policing, 41–42, 58, 205–6
Ramón Arellano Felíx criminal organization, 86
Ramphal, Shridath, 125
ransoms and kidnappings, 100
Raudenbush, Stephen, 52, 55
Raymond, Catherine Zara, 180, 182, 184, 192, 197
Regional Cooperation Agreement on Combating Piracy and Armed Robbery against Ships in Asia (ReCAAP), 182, 186, 188, 190, 249n92
Regional Maritime Security Initiative (RMSI), 176, 189
remittance networks, 78, 120, 146
repatriation, 106–7
Revolutionary Armed Forces of Colombia (FARC), 113–15, 148
Revolutionary United Front (Sierra Leone), 151
Rice, Condoleezza, 195
Richardson, John, 31, 33
RMSI. *See* Regional Maritime Security Initiative
Roach, Ashley, 179, 181–82
Rosenberg, David, 178
Royal Bahamas Defence Force, 76
Royal Canadian Mounted Police, 85, 130
Russia: as hybrid threat, 17–18; narco-sub collaboration with Columbia, 136
Rutledge, Clifford, *116*

SALW. *See* small arms and light weapons
Sampson, Robert, 52, 55
San Pedro Sula, 92
Sandiford, Lloyd Erskine, 67
São Paulo, as feral city, 16
sea power strategy. *See* maritime security; naval strategy
security. *See* maritime security; naval strategy
Seelke, Clare, 99, 101, 108, 110
self-propelled semisubmersibles (SPSS), 77. *See also* narco-submarines
seminary experiment, 38, 51
September 11, 2001, terrorist attacks, 30, 68, 181, 192; strategic impact of, 30, 68, 181
sequestration. *See* budgets and sequestration
sex trafficking. *See* human trafficking
sexually transmitted infections, 50–51, 101
Shower Posse syndicate, 122–24
Shy, John, 21–22
Sierra Leone: piracy in, 163, 164; Revolutionary United Front, 151; trafficking in, 151
signal crimes, 59
signals of disorder. *See* disorder
Silver, Nate: *The Signal and the Noise*, 54–55
Simon, Sheldon, 182, 183, 185, 187–88, 190, 191, 200, 249n92
Sinaloa Cartel, 72, 73
Singapore: drug trade in, 178; geopolitics of, 176–77, 179; maritime security in, 182–83; piracy and, 181, 185–86, 190; terrorism and, 193, 194
Singapore Strait. *See* Straits of Malacca and Singapore
Sittnick, Tammy, 195
Skogan, Wesley, 53, 220n9, 221n54
slums: broken windows theory and, 45; in Liberia, 157; as security threat, 1–2, 14–15, 206. *See also* urbanization
small arms and light weapons (SALW), 151

smuggling. *See* drug trade; trafficking
smurfs, 120
social spaces in police operations, 20
soft power, 16
Solanki, Amar Narayna, 1–2
Somali piracy, 160–62, 164, 171, 187
Song, Yann-huei, 182, 183, 189, 195, 196
Sousa, William, 45, 48, 53
Southern Command, 83–85, 108, 129
Southern Partnership Station 2017 (SPS-EPF 117), 88, *116*
Spearhead (T-EPF 1), *88*, *140*
speedboats and piracy, 165
Spratly Islands, 177, 179
SPSS. *See* self-propelled semisubmersibles
"squeegeeing," 46–47, 67
St. Kitts and Nevis: corruption in, 125; drug trade in, 79, 91; investor visas and, 111; money laundering in, 119; security initiatives in, 93
St. Lucia, human trafficking in, 100, 124
St. Vincent, drug trade in, 76
Stanford Prison experiment (1971), 57
Starr, Stephen, 166
State Department: on money laundering, 121; *Patterns of Global Terrorism 2002*, 68, 112
Steffen, Dirk, 161, 164, 244n176
stop and frisk, 42
Straits of Malacca and Singapore: broken windows theory on, 175, 177, 180, 192–203; counterpiracy initiatives in, 181–85, 191; drug trade in, 178; geopolitics and commerce of, 175–80, 245–46n6; human trafficking in, 179; irregular migration in, 179; Malacca Straits Sea Patrols, 182; money laundering in, 178; piracy in, 179–92, 197; strategic context for, 8–9; terrorism and, 193–95
students and school disorder, 51–52, 118
Sullivan, John, 15, 17, 20
Superferry 14, 2004 terror attack on, 194
Suriname, 72
Swartz, Peter, 29
Swift (HSV 2), 170

Talsma, Lara, 100, 103, 106
Taylor, Charles, 147
Teo Chee Hean, 194
Tepp, Eero, 157–59, 161, 164, 172
terrorism: Caribbean Basin and, 109–15, 117; feral cities and, 17; illicit trafficking and, 26, 68, 109–15, 120–21, 162–63; international treaties on, 29; piracy and, 166–67, 192–96; strategic response to, 25; white papers on, 26. *See also* counterterrorism; *specific attacks*
Thachuk, Kimberley, 123
Thoumi, Francisco, 150
threat-specific approach, 85
Thunder (Nigeria), 148
Tidd, Kurt, 82, 83, 103, 109
Till, Geoffrey, 4, 23, 24, 27, 59, 149
Titan, Operation (2008), 115
Tivoli Gardens, Jamaica, 122–24
Togo: drug trade in, 145, 147; gun trade in, 151; human trafficking in, 155; money laundering in, 152; piracy in, 164
tourism, 67, 100, 107–8, 190
TPD. *See* Transit Authority Police Department, 42, 44
Tracy, United States v. (2011), 110
trafficking: of arms, 91–96, 151, 155; of cash, 119–21; deportations and, 106–7; human, 96–102, 106–8, 110, 126, 152, 155–57, 179, 230n51; irregular migration and, 98, 102–7, 117, 155–57, 179; methods of, by sea, 75–77, 90, 120, 145, 146; naval literature on, 27, 31–32; public health risks of, 101, 105, 156–57, 179; remittance networks and, 78, 120; strategic responses to, 28–29, 127–28; West African instability and, 150–53; white papers on, 26. *See also* counternarcotics; drug trade
Transit Authority Police Department (TPD), 42, 44
transnational organizations. *See* nonstate actors and transnational crime organizations

transportistas (trafficking groups), 92
Treasury Department, 83–84
Trelawny, Chris, 214
Trinidad and Tobago: disorder in, 125; drug trade in, 75, 76, 124; human trafficking in, 100
Tuvalu, 119

UN Convention on the Law of the Sea (UNCLOS), 160, 161, 247n31
UN International Narcotics Control Board, 69
UN Office on Drugs and Crime (UNODC), 169; *Global Trafficking in Persons Report* (2012), 97, 98
UN Security Council, 169
Unified Resolve, Operation (2012), 83
United Self-Defense Forces of Colombia (AUC), 113
United States: Caribbean Basin and, 68, 93–96, 108, 117; deportations from, 106–7; as drug market, 69–72, 79; drug smuggling to, 70–71; foreign policy, 22, 68–69; Gulf of Guinea and, 145, 168; humanitarian assistance from, 198; investment in Latin America, 107; investment in war on drugs, 86; littoral areas' threat to, 6; migration to, 102–3; MS-13 in, 106; mutual legal assistance treaties and, 85; Straits of Malacca and Singapore and, 181, 183, 189, 195; white papers, 14, 25–26
urbanization, 14–16, 26, 29, 31, 34, 45, 143
urbicide, 20, 94, 122–23, 162
Uribe, Alvaro, 91

Valencia, Mark, 180, 193, 195, 199, 200
Venezuela: deportation and, 107; drug trade in, 74, 76, 81, 91; human trafficking in, 98–100
vigilantes and vigilantism, 125, 150–51, 154–55, 241n94
violence: in feral cities, 15–17; homicide rates in Caribbean, 91–96. *See also* guns and gun-related crimes
Virgin Islands, U.S., 124–25
Vreÿ, Francois, 143, 147, 153, 161

Waesche (WMSL 751), 199
war on drugs, 86. *See also* counternarcotics
Wave Knight (UK), 84
weapons smugglings. *See* arms trafficking
Weber, Max, 17
West Africa. *See* Gulf of Guinea and West Africa; *specific countries*
West Africa Cooperative Security Initiative, 148
White, Hugh, 24
white papers, 14, 25
Wilson, James, 38–43, 45–50
Windward Passage, 69
Winter, Donald, 33

Yar'Adua, Umaru, 168
Young, Adam, 180, 193, 195, 199, 200
Young, Mary Alice, 79, 122
Yucatán Channel, 69

Zackrison, James, 70, 71, 75, 79, 86, 119, 120, 135, 136
Zephyr (PC 8), *64*
Zimbardo, Philip, 39, 57
Zukunft, Paul F., 82
Zumwalt-class destroyers, 31

About the Author

JOSHUA TALLIS is an analyst at the Center for Naval Analyses. He has provided analytic support afloat to naval forces in the European and Central Command theaters and has contributed to analyses for AFRICOM, SOUTHCOM, and the Office of the Chief of Naval Operations. Previously, Tallis was manager for research and analysis at a security services firm in Northern Virginia. He holds a PhD in international relations from the University of St. Andrews and a bachelor's degree in Middle East studies from the George Washington University.

The Naval Institute Press is the book-publishing arm of the U.S. Naval Institute, a private, nonprofit, membership society for sea service professionals and others who share an interest in naval and maritime affairs. Established in 1873 at the U.S. Naval Academy in Annapolis, Maryland, where its offices remain today, the Naval Institute has members worldwide.

Members of the Naval Institute support the education programs of the society and receive the influential monthly magazine *Proceedings* or the colorful bimonthly magazine *Naval History* and discounts on fine nautical prints and on ship and aircraft photos. They also have access to the transcripts of the Institute's Oral History Program and get discounted admission to any of the Institute-sponsored seminars offered around the country.

The Naval Institute's book-publishing program, begun in 1898 with basic guides to naval practices, has broadened its scope to include books of more general interest. Now the Naval Institute Press publishes about seventy titles each year, ranging from how-to books on boating and navigation to battle histories, biographies, ship and aircraft guides, and novels. Institute members receive significant discounts on the Press' more than eight hundred books in print.

Full-time students are eligible for special half-price membership rates. Life memberships are also available.

For a free catalog describing Naval Institute Press books currently available, and for further information about joining the U.S. Naval Institute, please write to:

Member Services
U.S. Naval Institute
291 Wood Road
Annapolis, MD 21402-5034
Telephone: (800) 233-8764
Fax: (410) 571-1703
Web address: www.usni.org